Wild Hunger

Wild Hunger

The Primal Roots
of Modern Addiction

❧ BRUCE WILSHIRE ❧

ROWMAN & LITTLEFIELD PUBLISHERS, INC.
Lanham ✦ *Boulder* ✦ *New York* ✦ *Oxford*

ROWMAN & LITTLEFIELD PUBLISHERS, INC.

Published in the United States of America
by Rowman & Littlefield Publishers, Inc.
4720 Boston Way, Lanham, Maryland 20706

12 Hid's Copse Road
Cumnor Hill, Oxford OX2 9JJ, England

British Library Cataloguing in Publication Information Available

Library of Congress Cataloging-in-Publication Data

Wilshire, Bruce W.
    Wild hunger : the primal roots of modern addiction / Bruce
Wilshire.
        p.   cm.
    Includes bibliographical references and index.
    ISBN 0-8476-8967-0 (cloth : alk. paper)
    1. Dependence (Psychology).   2. Compulsive behavior—Psychological
aspects.   3. Senses and sensation.   4. Human behavior.   5. Substance
abuse—Psychological aspects.   I. Title.
RC569.5D47W55   1998
616.86—dc21   98-3356
                                                                        CIP

Design by Deborah Clark

Printed in the United States of America

♾ ™ The paper used in this publication meets the minimum requirements of
American National Standard for Information Sciences—Permanence of Paper
for Printed Library Materials, ANSI Z39.48–1984.

❧ REBEKAH ☙

Him wild hunger drives o're the beauteous earth.

—HOMER

The one thing we seek with insatiable desire is to forget ourselves, to be surprised out of our propriety . . . to do something without knowing how or why . . . Nothing great was ever achieved without enthusiasm. The way of life is wonderful; it is by abandonment. . . . Dreams and drunkenness, the use of opium and alcohol are the semblance and counterfeit of this oracular genius, and hence their dangerous attraction for men. For the like reason they ask the aid of wild passions, as in gaming and war, to ape in some manner these flames and generosities of the heart.

—EMERSON

In the long run . . . humans cannot tolerate ecstasy deprivation.

—FELICITAS GOODMAN

Contents

꧁꧂

Hunger for

Ecstatic Connectedness

My very chains and I grew friends,
So much long communion tends
To make us what we are . . .

—BYRON, "THE PRISONER OF CHILLON"

Waking Up
I was working on a draft of this book in the upstairs of our home in Plainfield,
New Jersey. The town lies within the urban complex surrounding New York
City, about twenty-five miles from Manhattan. Happening to glance down into
the backyard, I saw something odd. I put on my glasses.

A hawk or large falcon—mottled brown and white, standing erect and
alert—an inert pigeon lying, purple breast up, between its great clawed feet!
The raptor scanned the scene with its golden eyes, then with rapid movements of
beak plucked feathers from the pigeon's breast. Its watching ceased momentarily.
A patch of skin exposed, it tore off small strips of flesh. Instantly the scanning
resumed as the bird swallowed the meat.

I grabbed binoculars and spotted it, but it was too close to be focused. Backing
up, flattening against the wall, adjusting the binoculars to their near-range
limit, I got one clear image shrouded by fainter ones. Glued to the bird's beak
was a small bloody feather. Its eye—a great reticulated yellowish orb—shone
like living glass.

Then the visual field exploded. I lowered binoculars to glimpse the bird flying
up through the trees to the left. Near the fence to the right was the neighbor's cat
approaching the pigeon.

Not wanting contact with the wild creature to end, I banged on the window and yelled to scare the cat away. During the fleeting moments of the bird's visit, I had felt intensely—strangely—awake.

When I think of this incident some alertness and excitement return—and questions. Why was it strange to feel awake on a beautiful summer morning? Why did it take a wild bird's visit to arouse me? And from what? If someone had asked a few moments before the bird brought its lunch into view, I would have said I was wide awake and functioning on a high level.

I saw that I was not really awake. I live in a one-family house with locks and tight-seal storm windows, and travel strapped in a car. My feet, encased during waking hours in leather and rubber, are rarely bare and seldom touch the earth. If something jostles, I focus for a moment, then lapse into semisomnambulism.

I saw that I habitually perceived things only to the extent of tagging them for routine, everyday needs. If they didn't answer to these they were almost totally ignored. There was a smear of them somewhere vaguely "out there," a gray film that carried the subliminal message, DO NOT ENTER. Joyful possibilities of exploration were not imagined. Missing was any sense that anything was missing.

Addictions and Loss of Primal Experience

Occasional glimpses that something has been missing are insufficient. "We walk in worlds unrealized," Wordsworth wrote. But how to locate what we usually have no sense we are missing?

Human life was formed through millions of years in which our human and prehuman ancestors survived only by coping with wild Nature. By the Paleolithic era, humans made sense of wilderness in story, art, religion, and primal technology. Even when terrified at times, they probably did not feel emotionally empty. I strongly suspect that on one level we still hunger for primal excitement, but the hunger is partially suppressed and confused by overlayings of later agricultural, industrial, and now electronic life.

These days humankind's relationship to wilderness is strained and ambiguous. I think that *addictions* stem from breaking the participatory bond our species has had with regenerative source, with wild Nature over the ages—kinship with plants and animals, with rocks, trees, and horizons. Even terror is a bond with what terrifies. In such moments we are "out of ourselves," ecstatic, spontaneous, full of the swelling presences of things. Addictions try to fill the emptiness left by the loss of ecstatic kinship. They are substitute gratifications that cannot last for long—slavishly repeated attempts to keep the emptiness at bay. Finally they drain the body of its regenerative powers.

Civilization and Wilderness

It is easy to think that civilization should obliterate wilderness. But our bodies still cycle in the ancient regenerative rhythms of the demanding and supporting Nature that formed us and our prehuman ancestors over untold millennia (Nature is supporting if we are respectful, skillful, and fortunate). For example, getting a good night's sleep after the exertions of the day is replenishing. When a day's cycle aborts and cannot complete itself because of time changes in a long plane flight, we get "jet lag." The cycles of the body are disentrained from the cycles of the regenerative Earth that formed us.

But we needn't fly to be disentrained and drained. Our twenty-four-hour electronic society rides roughshod over regenerative rhythms built into our bodies, as well as those of all other diurnal mammals. Many scientists now maintain that most of the population is sleep-deprived, and that artificial stimulants are used to maintain a feverish semblance of wakefulness.[1] If these are withdrawn, withdrawal sufferings set in. Basic rhythms never let go of us, though they may be suppressed or jangled.

Physiologists show that the brain, as it has evolved over millions of years, generates its own "uppers," the endorphins and analgesics. These are natural opiates needed for survival in wilderness—rewards for meeting the demands of Nature and satisfying basic periodic needs. Lacking a minimal level of these opiates—no less now than then—we fall into "ecstasy deprivation," as Felicitas Goodman says.

If this deficit isn't filled in one way, it will be in countless others, at least by those who don't give up completely and wither away. Addiction takes many forms: drug taking and smoking, obsessive shopping or compulsive talking, or perhaps systematic eating and disgorging, the body punishing itself in a strangely ecstatic way, a version of ascetics' orgiastic mutilations of the flesh. Twisted ecstatic dependencies mimic and destroy regenerative involvements in ongoing Nature.

Today, in relatively safe environments, we can ingest euphoric drugs, say, whenever we wish, repeating ecstasies out of sync with ever-cycling regenerative Nature. But this impairs or destroys the regenerative capacities, the opioid-producing capacities, of our own brains. Addicts have lost their support in Nature. Dependent, prey to withdrawal anguish, they must prop themselves up perpetually.

Primal Needs

There are survival needs at least as old as our hunter-gatherer ancestors. Some are much older, those we share with other animate beings: needs for air, water, food, minimal shelter, and for sensory variety and richness. Then the more distinctly human needs, roughly these: for steady affection and responsive support during a prolonged infancy, and for confirming initiations through stages of development beyond this; for exploratory movement both early and late, and for a wide variety of stimuli to incite and reward it; for induction into respected stations and roles in the human group, and for sharing in the group's placing of itself in the World-whole, its origin stories. And then, despite deep cultural and cosmical identification, a need to develop what each of us is in addition, an individual being (which varies considerably from culture to culture). Sensuous and graphic—frequently playful—satisfaction of these needs is intrinsically valuable and ecstatic, and imparts to us a conviction of our significance and vitality as persons.

Taken as a whole, our primal need as bodily selves is to *be* fully through time, in cycles of exertion and restoration, probing exploration and recollection: to progressively discover our being in the wide world, not just to have it, as if it were a possession.

So insistent and powerful is the need to find one's life significant that the satisfaction of other primal needs may be sacrificed to fill it. Even life itself may be sacrificed by some to achieve a feeling of belonging and contributing.

The Stuck Alarm Bell

If primal needs have not been satisfied at an early age, we have no clear idea of what they are and what could satisfy them. That is, habitually unmet needs, constant and contrastless, appear normal after awhile, and recede into the completely taken for granted in the hazy margins of consciousness. Or their residual painfulness prompts their suppression from focal awareness. We suffer ecstasy deprivation nevertheless. There is the ring of a stuck and unlocatable alarm bell—we move here and there trying to find it, or we plug our ears.

Because these unmet primal needs still exist, a signal of alarm at *something* impending—starvation, collapse, ostracism, vacuity, whatever—will continue to sound, muffled and unplaceable, creating a vague, pervasive restlessness. We hunger for the wild, but "shoot wildly"—erratically, impulsively—trying to satisfy our hunger. This is addiction.

The symptoms may not be obviously nervous or flailing repetitions to quell the disturbance. If they were always so, addiction would be more easily identified and the needs it tries to fill would, perhaps, be more easily met. In fact, addiction is typically more subtle and insidious, for in many cases the alarm can be quieted quite easily, and we can appear at least to be calm. Psychiatrists call this "adaptation." That is, some repeated substitute gratification—it may be coffee, it may be cocaine—raises the threshold for hearing the alarm, and it is no longer heard. The body is forced into artificial equilibrium. Its organs are stressed, and vitamin, mineral, and electrochemical reserves taxed. But as long as these hold out, and substitute gratification is available, and calls from the larger self-regenerating world go unheard, all may seem to be well.

The consequences, however, are cumulative and destructive. The body's systems have taken shape through countless millennia of regenerative adaptation within Nature. The continued stress of addictive "adaptation" results ultimately in obvious maladaptation, degeneration, and suffering. To maintain the artificial equilibrium—and believing that no other is available—we prize and crave the very thing that damages.

There has probably never been a culture completely free of addiction. Settled fecklessness is perhaps universal in a certain percentage of persons at any time and place. The need to have one's significance recognized goes unmet and addiction may result. Or pain may be chronic—how about aching teeth?—and the need to dull it addictively may take over.

But why, with our wealth and scientific-technological powers of knowledge and control (pain-killers, for example), are we so limited in controlling addiction? The roots of our needs and the sources of our lacks and pains must go very deep. It must be that not all primal needs are easily identified. How about our need to find excited significance in things around us, an excitement that makes us eager to live each day—excitement that vouches for itself, is self-evident and unconditional in value, that confirms our own significance? Why do so many lead flat, addicted lives?

The current plague of addictions suggests a pervasive fear and distrust of the world and of our own organisms, a failure to stay awake and be vulnerable, to ecstatically find meaning over time. Only the willingness to allow the stresses and pains of consciousness also allows vital elations of belonging and growth. Lacking this willingness or faith, terrified of suffering and of fear itself, we may find the temptation to go childishly for rewards without the pain and work too strong to resist. This is the trance of addiction.

If something is missing in our meaning-making, we typically lack the means to make *that* clear. Missing is any sense that anything is missing. A restless somnambulism reigns.

But Don't Many Today Get along Well Enough?

We live at the time of the greatest migration in human history: from rural to urban environments. This generates consequences beyond our ability to fully comprehend. Simultaneously, huge national and transnational corporations emerge. As we will see more fully below, employees often feel confined and tiny, unable to identify with the whole enterprise, unable to meet primal needs of exploration and feeling themselves to be significant.

In general, relative to other animals, we are extremely adaptable. Frans de Waal writes that we are a "species of born adaptation artists."[2] Nevertheless, there are limits to our adaptability, though they are ill-defined. If these are exceeded, we wind up sooner or later in "the behavioral sink," our immune systems damaged, our regenerative capacities impaired. We develop manifold allergies, for example.

If asked whether they strongly prefer big city and big corporate life, many will affirm that they do. But satisfying preferences is not identical to satisfying needs. Widespread addiction suggests that many who appear to adapt in fact experience underlying malaise: numbed restlessness, unrecognized or poorly recognized needs. Tranquilizers, continual distractions, multitudinous therapies abound. Persons on TV looking at us, it seems, but not seeing us—images flashing violently—all arouse the vague feeling we are being manipulated, not addressed—erased. Arrays of technology and commercial expansion bristling around corporations and various groups . . . some maintain the concept of addiction must be enlarged to include the society's. It is likely that much that appears to be adaptive is addictive "adaptation."

But why not simply ask, Are people in touch with Nature less addicted than are we? If they aren't, why all the fuss about our loss of attunement to Nature? But those in touch with the Nature that formed us over most of our species' development—Paleolithic hunter-gatherers in wilderness—have nearly been exterminated. We cannot know if widespread addiction occurred then (though it seems unlikely that addicted people in such groups could have lived for long). We can only try to piece together through patient experience—and whatever help we can get from science, art, and religion—a new balance within Nature. We must grope, for I think we lack even a clear ideal of what a fulfilled self is.

Dividing Us into Minds and Bodies: A Hopeless Approach to Addictions

No attempt to divide us into mind and body—or into mind and body and spirit—and then to add it all up to get our wholeness can succeed. This entranced conceptual chopping and reconstruction is somnambulism appearing to be wakefulness. I believe we *are* our bodies, *are* bodyselves meant to be ecstatically involved in the world.

William Burroughs: "Junk is not a kick. It is a way of life." If we take this literally—junk a way of life—we grasp how desperately addicts fight to preserve their addictions. They feel the only alternative is withdrawal sufferings and death.

Everything shrinks in addiction—will, imagination, criteria of judgment, body-self. In a trance only the next fix is imagined. One is not paced for ecstatic life in the fullness of time but must have ecstasy now. In this shrunken world, one's hunger is all-encompassing, overwhelming. Becky Thompson titles her book about addictive eating disorders, *A Hunger So Wide and So Deep*.

When hunger is defined in our scientistic age, it is typically in narrow and "hard" physiological terms as a problem of basic nutrition. So understood, it has been solved for the vast majority in North Atlantic culture to a degree probably unprecedented in history. And yet many suffer from addictive eating disorders. We should enlarge what we mean by hunger and what can count as satisfaction. And we should question whether hunger in the larger human sense can be defined at all.

When primal needs are unsatisfied, self is untrue to itself. It is bought off by substitute needs and ephemeral ecstasies.

How is ignorance of one's own true need and happiness possible?

Notes

1. For a compendious survey and start, see *The New York Times Sunday Magazine*, January 5, 1997.
2. *Good Natured*, 201–203. Full citations for sources referenced in the notes can be found in the Sources Section at the end of this volume.

PART I

Nature's Regenerative Cycles

Once again I had that . . . feeling of being *on the mesa* . . . —it was like breathing the sun, breathing the colour of the sky.

—WILLA CATHER

Every weary one seeking with damaged instinct the high founts of nature, when he chances into . . . the mountains . . . if not too far gone in civilization, will ask, Whence comes? What is the secret of the mysterious enjoyment felt here— the strange calm, the divine frenzy?

—JOHN MUIR

I hear the sound of Heywood's Brook falling into Fair Haven Pond— inexpressibly refreshing to my senses—it seems to flow through my very bones—I hear it with insatiable thirst—It allays some sandy heat in me—It affects my circulations—methinks my arteries have sympathy with it. What is it I hear but the pure water falls within me in the circulation of my blood—the streams that fall into my heart?

—HENRY DAVID THOREAU

Chapter One

Ecstasy Deprivation

and Addictive "Remedies"

Each Aboriginal group in Australia roots and identifies itself with its ancestor-source—an animal, insect, bird, reptile, whatever. As the group names itself, so it is. This ancestor-source is commonly translated into English as the group's Dreaming. *To disclose this naming and this knowledge to nonmembers is to imperil the order of creation and the very being of the group. Individuals live because the group lives within the ever present ancestor-source—honey-ant, white-breasted sea eagle, monitor lizard, and so on. The ancestor's story, indeed its ever present career, is simultaneously the people's, their life's blood, their ligaments and bones.*

Our first contact with full-blooded Aborigines was at the roadside store at the intersection of the Stuart Highway and the road to Uluru (Ayer's Rock) in the Northern Territory at the heart of the continent. We had been told that not a single Aboriginal group kept to its age-old ways of living in, with, and from the land. Perhaps this is true. But certainly all the Aborigines we saw were garbed in European dress and seemed dependent on foreign economies.

And yet they approached that roadside store as if from another world. About five—they may have been together—converged on the store barefooted over reddish, dusty ground. If asked, "Were they walking on the ground, supported by the Earth?" I would have had to answer Yes. But the immediate impression was of beings who belonged to Earth so intimately that they could float slightly above it, and at any moment move effortlessly in any direction. Though erect, they moved over the terrain like weightless serpents. I gaped at them, and they looked across at me from a vast distance. Did I seem as foreign to them as they to me? Were they in touch with something I could not clearly grasp but perhaps might, and should?

Today in late twentieth century North Atlantic culture we are masters of engineering, the beneficiaries of manifold technologies for achieving things desired, self-advertising paragons of management and control. Many problems yield before technology: eliminating plaque on the teeth, innoculating against polio, putting persons on the moon, etc.

But many people feel dissatisfied and needy. Their lives are flat. They experience ecstasy deprivation and very often addiction. Various technologies are used to attack the problem, their multitude testifying to widespread dissatisfaction over their effectiveness.

We should ask if science and technology can accomplish the first task, which is to describe clearly what the problem is. Now, the description of a problem must tie in to what would count as a solution, if that were to be found. For example, a condition of thirst can be specified as a problem only because we know that it is water or some potable liquid that, if found, would solve the problem.

But what are we to say of ecstasy deprivation? Thinking within the problem-solving model, we automatically assume the solution is ecstasy itself. But what is that? We cannot describe it decisively as we can the potable liquid that satisfies thirst, or the regular heartbeat that solves the problem of fibrillation, or the rockets that put people on the moon, and so on. The idea of ecstasy is open-ended, enticing, and at the same time exceedingly opaque.

I *thought* I was living fully and happily—that I was writing freely and well—on that beautiful summer morning. But only when the wild bird arrived did I become aware of how limited I had been. My life expanded abruptly. What might I be missing right now, working and reworking this paragraph? It must be something like what the wild bird supplied.

The inability to adequately describe ecstasy deficit compounds the difficulty, and exacerbates a constant and contrastless restlessness. Ever-repeated attempts to fill the deficit constitute the addictions with which our culture is rife. Even "feeling good" may not be good enough, and may prompt lives of quiet desperation, as Thoreau put it, or even the loud and angry kind.

Psychiatrists, trained in the scientific model of medical materialism, have coined a term, *dysthymia*. It is supposed to mean lack of enjoyment in one's life. Like technologists generally, they assume that the enjoyment that, if found, would fill the need and solve the problem can be described. But can it? In some cases it seems clear enough—for instance, in the case of gross sexual dysfunction, which is solved when a successful "performance" occurs.

But perhaps even here the solution is not clear enough. Some who proceed right on schedule to climax still search for something more. How about an anorexic person who no longer feels an uncontrollable urge to diet and is manifestly happier? That is clearer.

Yet in many cases what would fill the need is not clear at all. As noted, some people feel a lack in their very enjoyments. They may experience pleasurable sensations, but accompanied by negative emotions or moods. The technical term *dysthymia* suggests a solution to a describable problem, but it misleads. Our task is not to calculate and explain what solves the ecstasy "problem," but to describe what happens when the experienced world opens up, often completely unpredictably, and our lives feel whole and swell along with it ecstatically.

Who Are We and What Do We Really Want and Need?

I think we are not duplex creatures, minds or souls somehow attached to bodies, but are body-selves. Ecstasy that is regenerative and not just momentary is a bodily and total experience, not just a mental or spiritual one, and, just as truly, not just a physical one. Body-self is difficult to illuminate, but so is ecstasy deficit and what would fill it.

What is the whole self? There are tempting abstractions that oversimplify the difficulty from the start: for example, the old stand-by, "A sound mind in a sound body." Or, "The whole self is body and mind and spirit," or "body and mind and soul and spirit." As if we already knew what the enumerable components of self are, and all we have to do is to add them up. Presuming to be holistic, the assumption of enumerability reveals the frozen mind-set of mathematical science and technology, an approach to the world greatly profitable in some areas and wretchedly inadequate in others.

To understand the body-self we must have concrete examples of human conduct today, as its often bizarre tangle of needs press for satisfaction. Take these:

Item: The fear of wildness and wilderness cannot completely conceal in many of us an attraction to it. The feeling is uncanny and fascinating, attractive and fearful. Wilderness can terrify and *be-wilder*—that remnant remains in our language; but it can also excite and fascinate. Even risk of great loss is exciting for many: the greater the risk, the greater the excitement. As Paul Newman replied when asked about his car racing, "I race because nowhere else do I feel so alive."

Item: Certain wealthy housewives or socialites caught shop-lifting say they do it "for the rush it gives" (vestigial and confused hunter-gathering behavior?). The risk excites the adrenaline plus endorphin rush—they feel momentarily enlivened, refreshed, aware—and may be unable to give up stealing even if they choose to do so, for without it their lives are flat, pointless, empty. They are dependent on the next rush.

Item: A contemporary Anglo-American said, "When I was about fifteen, there was a large White Oak tree in a clearing in a woods fairly near our house. Like a magnet the tree drew me to it. I experienced the seasons through it. I felt protected sitting near it. When the farmer sold this land to developers I was worried, but thought they might spare it, build around it, for it should be the center of something. When they put an X on it, as they had on other trees they had cut down, I had my brother tie me to it. But they got the cops to arrest me for trespassing, and cut it down."

What needs do such bizarre behaviors—and others not so bizarre—attempt to meet? In today's secularized and technologized world there are no commonly accepted, reasonable-sounding needs or motivations to explain such acts. It seems that a residuum of the prehistoric world in which prehumans and humans were formed over millions of years emerges, particularly strongly in some persons. A world in which people were either intensely and habitually involved and alert as whole selves or didn't survive to procreate others uninvolved like themselves. A world that was exciting and dangerous, one of close escapes or disasters, of rapt and astonished gratitude or despair, in which life was vivid and incredibly valuable, lived side by side with death. And a world in which we lived cooperatively with animal, vegetable, and arboreal kin or didn't survive. A residual memory of such a world would seem to best account for the boredom often felt by those who finally achieve the ease and security they think they want. But how we would "prove" all this is not easily known.

Archaic Experience, Rituals, Signs, Symbols, Myths
Examples like those above at least remind us that we did not appear yesterday out of nowhere. Our prehuman and human ancestors lived a hunting and gathering existence in wilderness environments for 99 percent of genus Homo's existence.[1] It seems counterintuitive to suppose that our recent irruptive cultural and technical evolution has simply expunged the needs and capacities encoded in our bodies, some of it genetically, over many, many millennia.

Given us to ponder are the nearly inert remnants of what our ancestors have themselves already interpreted, their rituals, stories, legends, symbols, myths, signs—the medical logo, say, those intertwined serpents. Or the dim but real totems of sports teams or fraternal clubs—bears, eagles, cardinals, elks, rams, and so on. Or the various flags or insignia we fly and the atavistic clothes we often wear. Or the role that archaic initiation ceremonies still play in religious, sororal, or fraternal groups.

Today, outfitted with technological powers, many tend to assume we no longer need myths or rituals. We no longer need to remember what our great-grandparents remembered—"Yes, we were there, our small Mormon band on the banks of the Missouri that winter." We encase ourselves in controlled environments called buildings and cities. Strapped into machines, we speed from place to place whenever desired, typically knowing any particular place and its regenerative rhythms and prospects only slightly. But if we need to bodily know particular places, then a primal need is not being satisfied. Addictive substitute gratifications may seduce us.

How much have we really changed from our hunter-gatherer past? As body-selves can we feel at home in the world without a story habitually told to ourselves in enwombing and orienting home-places? A story that inducts us into the regeneratively cycling universe beyond our sporadic life in buildings and speeding machines, and seals our membership in gratitude and awe?

At times, interlaced myth seems still to fall over us today, a spidery, weirdly resilient network drawing us toward sources in wilderness. Frederick W. Turner:

> For moderns the experience of these archaic visions is simultaneously strange and strangely familiar, as if, reading or listening to an unknown narrative we should gradually become aware of a rhythm of events announcing itself in advance so that we foreknow the conclusion.[2]

Ecstasy

The roots of the word *ecstasy* extend deep into early languages and probably prehistory. In Greek *ek-stasis* means *a standing out* from the points in space one's body occupies. To stand out into the surrounding world and to be caught up and possessed by it. The world owns me and, in a strange sense, I own it. For instance: swimming confidently in great rolling waves, they are made mine in my very abandonment to them.

But being transported ecstatically is not necessarily pleasant. We may be seized by terror, unable to anticipate a likely turn of events that will make sense of what is happening. We cannot *pay* attention; our attention doesn't seize but is seized. Yet we hang on and endure. The terror is felt to be *mine*, an integral element of a cohesive self.

Neuroscientists report that certain sorts of stress are beneficial, for analgesic and opioid chemicals are released naturally and periodically in the brain—the euphoria-producing endorphins.[3] At such times people are not left disoriented in the slough of tedium and addiction that mires so many considered successful today. Mired because substances and behaviors that substitute for opioids released cyclically and naturally destroy the brain's ability to produce its own "uppers." When input ceases, the ravaged brain demands another dose to counteract depression or convulsion. Dependency sets in, and sense of self, power, and worth withers—quickly or over the long term.

Some writers divide substance addictions from behavioral ones, divide the ingestion of hallucinogenic drugs from workaholism, for example. Disclosing self to be body-self shows the distinction to be artificial. Both groups need ecstatic fulfillment, and both fail to trust Nature to provide this in the fullness of time: the natural necessity of periods of recuperation or dormancy is not respected. Both groups must perpetually prop themselves up and short-circuit regenerative Nature. As the "crack" addict does not stop taking the drug, the workaholic does not stop working. Not confident that their organism in their place will regenerate itself and themselves in time, cyclically, they are not in possession of themselves. They are, to various degrees, possessed.

Wilderness

The roots of the word *wilderness* convey its attractive/ frightening ambivalence, its uncanny power to excite frightened desire. At first glance they simply mean "wild place." But wild and self-willed or willfull are connected.[4] In this reconstruction, the roots of *wilderness* are *wil*, plus *der* (of the) and the Middle English *ness* which means place. Most revealingly, "wil-der-ness" connotes "the will of the place." Wilderness has its own periods and ways. Cultures since the advent of agriculture nine thousand years ago have striven to conquer the will-of-the-place. Yet we have continued, apparently, to long for the excitement of will-of-the-place that catches us up as a vital part of itself.

Spasms of orgasmic pleasure are only a part of what is desired in sexual experience. A new and attractive body is exciting just because it is new. Underneath the clothes lies an unknown wilderness to be explored—the will-of-the-body-place, a wilderness with its own will. What might happen when we abandon ourselves to each other? Essential to the excitement is often some stress or fear. Thus the quest for—or the nagging urge to quest for—new individuals to disrobe and explore.[5]

> License my roaving hands, and let them go,
> Before, behind, between, above, below.
> O my America my new-found-land . . .

So cunningly John Donne links discovering the woman's body and discovering America. But is the will-of-the-place, the woman's will, respected? Or is she being exploited, as was America? Perhaps sexual experiences are merely sexual conquests, and become so frequent and easy that satiation with that type of exploration ensues. Fear—meaning awe—is gone. In unjaded sexual activity we hope to become aboundingly real and whole through each other. If the full panoply of ecstasy is not found, and we have no clear idea of what is missing, we may feel compelled to repeat what we do find. Ecstasy deficit excites mechanical repetition, addiction. Gerald Sykes:

> For the city man the only green thing is sex. For the country man there are many green things. Freud was a city man, Jung a country man.[6]

What happens to a controlling man when women develop their own ecstatic possibilities? In a time of fluid roles and decaying patriarchal hierarchies, this is a dismaying prospect for many men.

Technological mechanics steps forward, offering its own ecstasies, which aim to overwhelm wilderness Nature's. For example, "interactive television." One film features a nubile porn actress. Various commands are available: "Take off your bra," "Touch yourself." Technologized sex is a vortex that magnetizes and constricts.

I was not happy with my life (others were not happy with it either). When I was driving, a car going too slowly ahead cut me off from an earth-shaking objective that somehow had to be visible through the tunnel of my line of sight. What this was precisely, I would not have been able to say. Fame and glory? Yes. But the hunger overflowed these, engulfing me in a vast, churning amorphousness. The poet Hölderlin's line, "It was no person you wanted, believe me, it was a world," pulsed in a cocoon, as if it might carry a cryptic answer. Some idiots were said to cry for the moon. How could a sane person want a world? My "tunneling"

*marked a desperate desire to plunge into a world that would welcome me into
itself only if I possessed skills and words of request I did not know.*

*There were other moments of feeling encompassed protectively, relaxed, as if
a voice were murmuring in the trees, the river, or the breathing sea. Vaguely
heard, as if promises had been made from earliest days, "I will take care of
you. You will be at home with me, knowing me." Muffled words, reverberating
somewhere beneath the threshold of ordinary speech. Who or what was speaking?
There was a blank. Though I was an adult, a professor, a husband, a father, I
could not, at will, redirect or stop the habitual tunneling plunging. The loss of
freedom was loss of myself.*

Regeneration

A key feature of wilderness is that any place (if not fatally poisoned) left
to itself, will regenerate some way in the fullness of time. The Nature
in which our forebears were integrated through seasonal rites and rituals in home places was perceived by them to be both repetitive and
regenerative. We are therefore predisposed to expect regeneration
through repetition of *any* of our acts. This is delusion.

For many thousands of years our ancestors survived only by fitting
into the self-regenerating whole of Nature that fed its dying parts back
into itself, to repeat the life cycle. Sprouting, blooming, maturing, fad-
ing into death, blooming again. Animals growing, rutting, birthing,
dying, and the cycle repeating itself. All things: birth, flourishing,
dying, rebirth. Nature feeds back its "wastes" into itself. The heavens
repeat the same perpetual cycle of waxing, waning, dying, reawakening
of celestial bodies. Humans enjoy excited consummations, dying away
of excitement, and typically its rebirth in the fullness of time. Human
rhythms—so-called biological clocks—evolved as cultures wove the
rhythms of humankind and all of Nature together in myth and ritual.

But many today lack assurance that the inevitable demise of excitement will be followed by its rebirth in the fullness of time. In fear of
emptiness and inertness, ecstasy must be mechanically reinduced. Addiction is failure to trust the spontaneous recoveries of Nature and culture.

Addictive gambling, for example, is repetition without orientation.
It mimics the spontaneities and risks of Nature herself, but also her
chances for fabulous abundance, a windfall. Convinced they can win
it all, a world somehow, gamblers blur, trancelike, the lines dividing
possibility from actuality and part from whole.

But only money and what money can buy is possible. One is penned up in addictive repetitions—like a wolf or coyote relentlessly pacing its cage—with a wild hunger and no way to consummate the longing.[7] How better explain such counterproductive behavior than supposing some badly identified primal need that drives the gambler? And that the concealed need is deep enough to conceal its own concealment and baffle attempts to satisfy it? Vincent van Gogh:

> We cannot always say what it is that surrounds us, imprisons us, and seems to bury us; and yet we do feel these indefinable barriers, these railings, even walls.[8]

Mother and Mother Nature

Before we came into the world, we were already in the world. As the unborn infant floats in the womb's amniotic fluid, the will-of-the-place floods its nervous system.[9] But so suffusingly, nurturingly that no sense, presumably, of self-over-against-a-world arises. Yet already the body's later cyclicity begins to form as the infant interdigitates with the world's cyclicity through the mother. Her moving about and uttering sounds irregularly, her various perturbations during the day, and her relative immobility at night, when the solemn beating of her heart and the gentle rocking rhythms of her breathing achieve their nocturnal dominion—world fulsome and abundant beyond reckoning.[10]

Birth is some kind of sundering of an ineffable self-world whole. Without the compensation of enwrapping support, particularly from the nursing mother, the infant is prey to abandonment terrors. How can we help supposing that if basic hungers are unmet in those days, the emerging self will clutch frantically at a world it cannot trust? Panicked attempts to minister to body-self amidst hopelessness are presumptive roots of addiction.

The enwrapping or looming mother is for each of us Nature's first reality, either greatly frightening or greatly comforting, or both at different times. She is also, of course, an enculturated being. "Mother Nature" is root metaphor, a likeness too close to experienced Nature itself for analysis to easily pick the two apart and too engulfing emotionally to be reduced to objectifying, literalizing terms. The mother of all metaphors? It emerges early as the background of all attempts to understand the hungering for connectedness, wholeness, growth, and the complex cultural strategies over the millennia to achieve these.

Guilt

Inability to realize our deepest needs and capacities for ecstatic wholeness is a failure of responsibility to ourselves, and engenders hazy, floating guilt. But guilt is noisome and must be suppressed. So we may "pop an upper." But there is more guilt when inevitably the down follows—or threatens to follow—the artificial high. Guilt can be suppressed only through another dose of the drug or the behavior, and on and on.

This explains the fateful progression of addictions in ever greater distraction and dependency, also the tenacious aura of mendacity that hangs about most of them, and why Alcoholics Anonymous insists on honesty. How can we admit to failing ourselves at the heart of ourselves?

The maintenance of addiction demands self-deception. Addictive "adaptation" requires that primal needs be suppressed, and the alarm at their going unmet be unheard. The exploratory urge to reach out into the self-regenerating world to contact other beings with whom we have evolved over millions of years is such a need. When these needs are suppressed, substitute gratifications take over. Primal needs are kept on the margins of consciousness: perceived as "not to be perceived further." We deceive ourselves.[11] Any trouble in our life is blamed on others. Or, trouble may be owned, but only to the extent of admitting lapses in managing gratifications: "I usually have drinks only with food, a couple of martinis at lunch, a few drinks before dinner, a little wine during, and a liqueur perhaps afterwards. But life has been so hectic lately I haven't been able to eat regularly." Self-deception encloses the addict in a substitute "world." It can be broken through only by bodyself in possession of itself, but this is what is lacking.

Addictive compulsions are irrational desires for things that, once attained, do not satisfy for long. As Eric Hoffer once put it in an interview, "You never get enough of what you don't really want." Even if we succeed in stopping, the possibility of resuming typically preoccupies us. The vast consensus of observers—including sometimes the addicts themselves—is that the addictive behavior impairs one's life. But it is difficult or impossible to stop it, for no longer possessing our ancient place in Nature's will-of-the-place we no longer possesses ourselves.

Civilization, Coordination, and the Sacred

Historically and prehistorically the most important task of civilization has been to survive by integrating us in will-of-the-place. Among earliest evidences of distinctly human cultures are lunar calendars incised on animal bones. That we might be caught up ecstatically as a vital part of the enwombing universe, its recurrent cycles were recorded and celebrated in ours—in rituals, ceremonies, stories, and myths.[12]

Vast mounds and circles of stones in what is today England and Ireland were erected around 3500 B.C. They registered the turning points in the progress of the Earth through the year, particularly its changing relationship to the sun: winter solstice, spring equinox, summer solstice, fall equinox. Stonehenge is a well-known example. These mounds and circles were not for merely practical purposes—planning for planting and harvest—but fed persons as wholes as they cycled with the seasons. The egg-shaped mounds and circles were temple-observatories.

Probably our early ancestors did not sharply divide activities into practical and spiritual. Each celebration recreated the whole history of the people by reinserting them yet again in the regenerating Whole. The most spiritual was also the most practical. The source that created them was never absent. Can we not suppose that, despite what we would call hardships, they were in possession of themselves, because in perpetual possession of their source and place in Nature, and that they lived in ecstatic kinship with things around them?

I imagine they experienced themselves belonging in what caught them up excitedly in a vast chorus of living and nonliving beings. At least Paleolithic art, much earlier than the stone circles, suggests this. Gathering-hunting people of our own century do not—or did not—define personhood in a way that isolated them from other beings, other "persons" (we put the term in scare quotes).

But this experience of belonging in what catches us up ecstatically is the experience of the sacred: of what roots and empowers us in the world and centers and orients us, of what we know can never be completely known or controlled. "Whole," "heal," "holy" are connected.

Loss of ego is a kind of sacrament that leaves room for something much greater than ego. "As it was in the beginning, so is it now, and ever more shall be, world without end, Amen." The Christian doxology exhibits roots in more ancient human practices of restoration and homecoming.

Fragmented Modern Humans

Despite tendencies inherited from our ancestors, the contrast between us and them (even fairly recent ones such as the Athenian Greeks) is vivid. As John Lachs, for one, points out in *The Relevance of Philosophy to Life*, we are caught up in vast organizations that use us as if we were cogs that click in and out at appointed times in vast institutional machines. Having little sense of the meaning and value of the Whole of which we are parts, we have little sense of our own meaning and value. Addictions are acts of violence directed at our own insignificance. Lachs writes:

> Our official roles require that we perform act-fragments without concern for their meaning or consequences. . . . [W]e soon learn to act . . . with little interest in how what we do fits current circumstances and with what its ultimate outcome may be. The psychic distance from the immediate and the long term context makes it difficult to introduce intelligent moral considerations into the deliberative process. . . . The fragmentation that pervades the mediated world is internalized . . . The well-known problems of coordinating intellect and feeling arise from the independent operation of what should be integrated personal functions.[13]

Though we may sneer at how small the world of indigenous persons is, still they had a visceral sense of the Whole as well as their own competence and worth within it. (Words referring to the unconditioned and unconditionally valuable have mythic force and are capitalized.) An anthropologist once spoke of a group of indigenous people who marveled at the drinking glass she owned. They asked how such a thing was made. When she told them she did not know, they did not believe her.

Knowledge must be rerooted in the ecstatic body-self that molds materials and becomes whole within the Whole. Henry Thoreau describes standing on the peak of Mt. Katahdin in Maine:

> What is this Titan that has possession of me? Talk of mysteries!—Think of our life in nature,—daily to be shown matter, to come in contact with it,—rocks, trees, wind on our cheeks! the *solid* earth! the *actual* world! the *common sense*! *Contact*! *Contact*! *Who* are we? *where* are we?[14]

From long experience working with addicts, Alcoholics Anonymous cites "belief in a higher power" as a first condition for recovery.[15] This seems to be true. Self-worth resides in a firm sense of the vast Whole in which we are small but vital parts. And the AA mantra or motto "one day at a time" suggests at least that our vitality depends on rootedness moment by moment in the sensuously given world around us, that sweetest gift that answers to our needs as bodies. Contact! Contact! Without this we can't be content with "mere Nature" but must contrive addictive gratifications.

But what exactly is "higher power"? If it means higher than the world, it will tend to disengage from the matrix of regenerative Nature.

It will also suggest fanatical allegiances, even if not to drink, commitments intolerant of the profusive variety, stately pace, and local integrities of the world. It will mean addictive short-circuiting at both the group and personal level.

Loneliness

Our Paleolithic ancestors almost seem to be another species. How to grasp their learning through millennia from the cycling lives of beasts, birds, snakes—what to eat, where to find shelter, when and where to hunt, gather and store, when to lie dormant?[16] Their quiet alertness is inconceivable for most of us.

Nevertheless, on some level of consciousness many of us still feel the will-of-the place, our alienation from it—our abandonment and loneliness. A member of the Omaha Tribe:

> When I was a youth the country was very beautiful. . . . In both the woodland and the prairie I could see the trails of many kinds of animals and could hear the cheerful songs of many kinds of birds. When I walked abroad, I could see many forms of life, beautiful living creatures which Wakanda [The Great Spirit] had placed here; and these were, after their manner, walking, flying, leaping, running, playing all about. . . . But now . . . sometimes I wake in the night, and I feel as though I should suffocate from the pressure of this awful . . . loneliness.[17]

In an age of power and technology, many feel powerless. Erich Fromm in *Escape from Freedom* wrote of the bewildering choices offered by modern technology. It destroyed the seamless mythic matrix that guided and empowered earlier persons' choices day by day, stage of life by stage and provided intrinsically satisfying excitements and contentments. Perhaps the very magnificence of sleek, powerful cars and the allure of the person beside one create disappointment when, still, there is something missing, we know not what.

An Addicted Culture?

Some studies maintain the whole culture is addicted.[18] As we throb constantly in robotized atemporality, the charged atmosphere appears normal and is hardly noticed. Robert Atkins, M.D., notes that our society consumes ten times as much sugar as any that has ever appeared on Earth.[19] This amount of sugar throws many into a headlong run, a perpetual attempt to try to regain balance and counteract withdrawal symptoms with yet another dose of sugar (or caffeine, or . . .). The

breathless atmosphere of NOW and the general commotion and amorphousness are all-pervasive—life in a dust cloud.

Scientific research tries to keep pace with the violation of the organism's ancient regenerative cycles, particularly that of sleeping and waking. With the abrupt emergence of the electronic age, artificial light
available twenty-four hours a day and the bombardment by messages is
relentless. Once the natural diurnal rhythms are overridden, a semblance of wakefulness must be produced by stimulants of various kinds,
which easily become addictive because their effects are ephemeral. If
we addictively do without sleep—a primal need much closer to other
diurnal mammals than previously suspected—we tend never to be fully
awake. The height of wakefulness is proportional to the depth and rip-
ened fullness of sleep.

Evidence suggests that a component of violence in many hardened
criminals is physiological, a matter of diet. Consumption of junk food
whenever hunger pangs occur leaves withdrawal symptoms so disturbing that sometimes only a violence-rush can allay them for awhile. Normal hunger pangs degenerate into addictive cravings. And we can be
addicted to the excitements of criminality. When offenders are put on
a farm and raise and eat vegetables, relapse into crime is reduced.[20]

How important is the physiological factor? Probably the attempt to
break down "the problem" into (presumptively) technologically manageable units or factors reflects the generally addicted society that
wants immediate solutions. For example, how could we differentiate
the chemical factor of the stimulant consumed (caffeine or sugar, say)
from the urge simply to consume anything to fill an experienced emptiness?[21]

All of us need to feel significant because we do significant things.
Mere consumption masks the need without satisfying it. In criminals
the need is explosively strong. In noncriminals the need is not necessarily vividly apparent. But a pervasive perturbation takes over. As long as
the gratifications demanded by the addictive "adaptation" are available, however, and our organism does not give out totally from the
strain, this need might not be obvious.

Hungry, crying, new-born infants do not know what will satisfy their
need, but satisfied, instantly feel better. Likewise, persons binging on
food cannot describe what they really need, but feel better while eating.
But the infant is satisfying a primal need and developing as an infant-
person, whereas bingers can no longer fill the dependency need and

are degenerating. Any ecstasy not coupled with growth and a sense of one's significance is degenerative.

In an addicted society, control and satisfaction must be achieved NOW. Units not relevant to immediate power and control are ignored. When these "irrelevant" units are the regenerative cycles of Nature themselves, there is reason for Thomas Berry's assertion that the culture exhibits a rage against the very conditions of life itself.[22]

Overreaching and Disintegrating Mind and Consciousness

Given the prevalence of addictions and vain attempts to escape them, our mental powers seem more entangled and weak than we commonly imagine. I do not think mind is separate from body (all of the body, of course). What, then, is "human mind"? How do we best deploy our "mental powers" to coordinate capacities of awareness and primal needs?

Other animals' exchange with their environment is regulated almost totally by instinctual turning of awareness and involvement. Wolf consciousness, say, is wolf-moose consciousness—that of predator-prey. Their hungers are naturally periodic. As long as these animals endure at all, they are regenerated by Nature and its balances and rhythms, which they cannot, typically, disrupt. They live in a vast womb.

For us, sounds and sights can be separated in thought from the things that emit sounds and show themselves. And it is just because these sounds and sights can be separated in thought from their things, that they come to symbolize and name them. They are *of* things that are absent, or that do not exist, but perhaps might.[23] We alter Earth in the light of our imagination and powers of abstraction as does no other animal. This is what many call the glory of human mind.

Yet what is the value of our distinctively ecstatic and far-reaching mind when it includes the capacity to distract ourselves from Earth's local environments cycling regeneratively (if not totally crushed)? What is the value of mind when it creates havoc with the human organism?

Body-Selves Creating Havoc with Themselves

A fundamental element of every local environment is our own organism at the center of it. But unlike the rest of that environment, it cannot be left behind. It is inescapable. Yet through the organism's constancy and contrastlessness we easily overlook it. This can be deadly. For when we mentally reach beyond the local environment, we tend also to overlook our primal needs as organisms to deal with local place and its rigors: to explore in it, achieve rich sensorial contact, and enjoy kinship with its quietly arboreal and bustling animal life.

If we lose this contact habitually, primal needs go unmet. We imagine immediate substitute gratifications—caffeine, cigarettes, cocaine, mere sex. We cannot wholeheartedly own up to these gratifications, for they are not needs that through evolution constitute our very being. They are counterfeits that lead to dependency and loss of self-respect.

When urges and desires are not willingly and happily appropriated as truly my own, they split off from the rest of body-self, particularly our will. They become peculiarly thing-y and "not me." But yet they cannot be experienced as merely physiological events either, for we *are* *body*-selves. So they are also bafflingly, mockingly "me."

This disorientation within body-self is a trance state in which addictions flourish. One more gratification, one more, one more . . . repetitions that isolate from the regenerative Whole. Alienation from the Nature that formed our needs over millions of years means alienation within ourselves. The easy opposition of mind to body—the ingrained distinction of analytic thinking—can never disclose how tightly and closely addictions hold us.

Neither ordinary hunger pangs nor addictive cravings can be stopped by simply commanding them. But the craving cannot be accepted wholeheartedly as mine, while the hunger pang can. Most suffering from bulimia would rather not crave to eat a third chocolate pie after vomiting the first two. They desire the pie, but do not desire to desire it. Typically they realize on some level that they violate themselves, and that this is degenerative. They cravingly desire it nevertheless—the craving "both mine and not mine"; one acts as if mechanically. Whereas anyone satisfying an hunger pang in the ordinary rhythms of stress and restoration can identify with it: "It is wholly mine, essential for my being and remaining myself." All this assuming that satisfying it doesn't violate the need to feel one's choices to be decisively important. For example, one may have obligated oneself to fast for a time.

The point of traditional cultures and their initiations into stages of development was to keep ecstatic needs and satisfactions integrated and the person coherent. If we are body-selves, then thinking that mind can stand off from body and manage it any way it wishes—that is going out of our minds.

A General Statement of Value

I think the greatest value is the greatest individuation within the greatest coherence and unity. What is intrinsically valuable for body-selves is that which strengthens and integrates them as individuals because it integrates them ecstatically within the world that formed us.

The danger is to lose integration within the world as Whole. Imagine the very worst: a nuclear event that darkens and freezes Earth for months or years and poisons it for centuries. There would still be the sun and the stars above the poisonous clouds. But since no sentient beings would remain, there would be no consciousness of light, only vibrations of energy interweaving through darkness unseen. Tehipite Dome would still raise its granite head four thousand feet above the King's River in the Sierras of California, but no birds would perch atop it, and no human or animal or bird would note streaks of yellow down its bluish face.

Indeed, it would not be a "dome," nor a "head," nor a "face," nor "Tehipite" (an indigenous name with a meaning we might appreciate). That is, without bird and animal consciousness it would not be seen, but without human consciousness it would not be named. And it would not cycle and web through vast regions of human appreciation in art, science, and religion. It would be a vast pile of minerals in space-time, some discovered and known by us at an earlier time, others undiscovered. The minerals once seen on its surface as yellow would be somewhat different in their molecular configuration from those once seen as blue, that is all. The excellence of our seeing, naming, interpreting would be missing; and the dome's excellence in being seen, named, interpreted would be missing also (though doubtless there is more to its excellence than this).

But our powers of consciousness and fabrication can equally well turn such places into a Disneyland. A huge fund of primal meaning and associations that links us with ages-old human and animal—real animal—consciousness can be obliterated. When images of animal kin are shrunken, we can only shrink in response and fracture within ourselves trancelike. Thus the question again: what is the overall value of mind and consciousness? It has a reach, complexity, and richness that can add to the richness of the universe. But we easily dislocate ourselves from real beings that populate our actual environment—real animals, not sleepless Mickey Mouses.

The Importance and Limits of Physiology

In 1932 the physiologist Walter Cannon published *The Wisdom of the Body*, showing how the very factors that throw the body out of homeostatic balance combine with the wisdom of the body to restore it. In 1977 the physician Vincent Dole published a short and pointed piece, "An Apparent Unwisdom of the Body," calling attention to the body's need for excitement: it will playfully or grimly throw itself off balance to generate the excitement of trying to reestablish adaptation or equilibrium.[24] Dole closed his piece with a call for "a new Cannon" to make sense of this phenomenon.

But our behavior cannot be understood in objectifying physiological terms alone, as essential as these are. A conceptual matrix that might equilibrate our lives cannot be framed solely in terms such as "homeostasis." Needed are "world-involvement," "ecstasy," "sources," "danger," "responsibility," "patience," "freedom"—terms that physiologists cannot employ in their integrity, but must break down and reduce. Above all, "self": my ability to bind experience over time as my own work, to be involved in my own sources, and, as I dip back into them, to stretch beyond them into a world that reclaims itself regeneratively out of its own sources. An ecstasy that is more than dependency on passing thrills.

About forty years ago I was fishing in the North Fork of the King's River. Finding the way around the tail of a large pool, and ducking beneath alders that hung branches low over the water, I walked bent over. The shade and dampness pulled me out of the noonday sun, away from the white granite of the canyon walls and the rocks of its floor that reflected light that I felt right through my hat, and certainly up beneath its visor.

Sitting on a rounded rock, I relished the extraordinary coolness. In a while a clicking sound emerged. There at the pool's edge hundreds of pieces of driftwood played, rubbing against each other in the miniature waves. At various stages of roundedness, whiteness, and polish, they told time with a patience and amplitude I could not precisely tell, only delight in. They had rubbed their substance back into the water and into the soil.

I picked up several pieces—one the abbreviated torso of a horse, another a serpent with upraised head—and put them still dripping in my creel. Unpacked when we left the mountains, they lay on shelves in the East for about thirty years, usually unnoticed. Then one day as I was looking at the horse, it became clear—for no reason I could give then—that that piece should go back where it came from.

When I again visited the North Fork, this time with my twenty-year-old son, I tried to find the pool, but could not. Perhaps the river had changed its course years back, and what was the pool's floor now lay baking in the sun. Perhaps in that place where water once shown in its depth, rippled at the shore, and alders once stood, black grasshoppers now flitted and snapped among the rocks.

I deposited the driftwood at the tail of a pool that bore a faint resemblance to the first. The horse now lay with pieces already there, clicking along with them.

Rooted Appreciation and Responsibility

Because we have great powers of appreciation, we have great responsibilities (The word has become dulled and leaden—as if responsibility could not itself be ecstatic). We feel other animals' rudimentary and powerful appreciation of things. We see a horse let out in pasture after weeks in the barn, its joy, or a dog's delight in anticipating and welcoming its owner. But it is highly unlikely that the animals can appreciate the World-whole as we can. We can grasp our present environment as a small part of our planet, and Earth as but a minute part of the ongoing universe. We can care about how our behavior affects the larger regions within which we nest. We can feel an awed attentiveness, a belonging and loyalty—a unique contribution.

To be responsible is to be responsive. Each thing is worthy of consideration in some minimal but fundamental sense. The merest bubble rising from the plashing of a waterfall is a primal achievement. Without the whole supportive matrix of Nature things that are clearly valuable wouldn't exist. This we can recognize even when we can't help destroying certain things.

We are beings with unrealized potentials for ecstasy. We don't have to argue for altruism or duty to Nature. We have to wake up. If really open to the world around us, if responsive, we will feel kinship and respect and obligation.

Some ecologists assume we must either take a "human-oriented approach to the environment," or a "thing-oriented" one, but not both. This is an artificial and deracinating either/or that belies our actual situatedness. For nonhuman things "in themselves" must be experienceable in principle and to some extent if we are to know what we mean by them. And to know what we mean by ourselves we must know how we would be experiencing them.

In other words, the facile phrase "things in themselves" has some limited but basic meaning for us because, for example, we have some idea of what it's like to run into a wall blindfolded. This is our sense of the brute reality of things—things as experienceable in this way. Nevertheless, it is *our* sense of it. On the primal level, then, we simultaneously grasp ourselves and the rest of the world—ourselves as meaning makers.

Without this ability to create meaning in terms of experienceability, how could we mean that some thing is *real*, that is, that it would exist even if there were no actual, direct experience of it? This enables us to value things for themselves even though they have no obvious and immediate use-value for us. But in this very lack they have immense value, fulfilling ecstatically our capacity to honor and protect, when we can, things for their own sakes.

"Ah, but isn't it limiting—anthropocentric—to think that everything must be experienceable in order to be referred to meaningfully?" No. Very probably there are sorts of things going on right now among us that are so strange we can't imagine them—can't imagine how they would be experienced if we could experience them, can't imagine to look for them. But then such things are experienceable even now as beyond our ability to imagine them, that is, they are *mysterious*; and that is just what we mean by such things being *real*. Which involves what we mean by *ourselves*: we are members of this mysterious world.

Emerson writes in "Circles," "Every man supposes himself not to be fully understood. . . . The last chamber, the last closet, he must feel was never opened; there is always a residuum unknown, unanalyzable. That is, every man believes he has a greater possibility."[25] This is the excitement of the mysterious, of what lies beyond our sight or grasp. Are the only alternatives to shoot wildly at possibilities that exceed all measure, or, on the other hand, to turn away from all possibility and wither away? I think Emerson suggests a third way: to find a fitting role within awesome "circular power returning into itself," a universe that creates, recreates, and also constrains itself. Such would fulfill the will and satisfy the heart. This book searches for that kind of meaningful involvement in a world in which technology and closely cribbed mediated life are givens. But we might cast them in a wider frame of acting and awareness, and dilate and respond to all of it.

We forget at our peril the consequences of this peculiarly restless and aggressive technologically equipped North Atlantic culture that crashes about, destroying species, for example, often without knowing they exist, destroying things marvelous before we can even imagine to look for them.[26] Conventional economic theory assumes that urges and desires are insatiable.[27] This is to suppose that there are no experiences satisfying in themselves, intrinsically and unconditionally valuable, that can curb, contour, round out desires and urges. But this is to assume a life running wild in the worst sense: desire for more desire for more desire, power for more power for more power, endlessly without satisfaction. Like a dynamo, ever accelerating, that tears itself to pieces.[28] Simply to assume that desires are insatiable is to accept as normal a mad form of corporate addictedness.

Smiles of Condescension

Traditional cultures inculcated a vivid, life-forming sense of the world as a whole and their members' vital place in it. Practices in some of these typically appall us, for example, human sacrifices to a sun-god. And examples of ignorance of processes that we think we know quite well with our science tend to bring smiles of condescension. But they did have a world and their excited membership in it.

There is some evidence that at least some individuals chosen to be sacrificed regarded this as the highest honor. The captain of the winning team in certain Mayan games was sacrificed. Again, in a south sea island a paradigmatic young couple engaging in ritual sexual intercourse was crushed by a roof of logs at the high point of passion. Beyond the excitement of feeling themselves to be members of a glorious, regenerating World-whole, they probably felt, in their very dying, ecstatically full of the people whom they represented, and the world for which they were sacrificed.[29]

We lack the words to even suggest this level of ecstatic fulfillment, and the whole business seems utterly grotesque to most of us who live in an "advanced" culture. Yet we still do root within the World-whole, whether we thematize it or not; we still are cosmical creatures, birthed and sustained by it. Outbreaks of mass-madness and the genocidal extermination of "others," as in Hitler's Germany or Pol Pot's Cambodia, or so recently, and perhaps again, in Bosnia and Rwanda, testify to forced and premature notions of wholeness. They bespeak the fragile corporate ego, the hysterical focusing of ecstatic energies.

Without proper channels for these energies, and lacking the excitement of even misdirected ones, many suffer profound loneliness and boredom. Recall Emerson's words in the opening epigraphs of this volume. Isn't the "oracular genius" that "craves to do something without knowing how or why," and lends the great danger to "opium and alcohol," a need to be caught up in the World-whole and to feel that exciting and spontaneous merging and flow? We seek a world-excitement that is, for us moderns, sane, a kind of Emerson-inspired third alternative.

Addiction As Loss of Self and Supportive World

Addiction is no easily formulable problem. It is a vexing difficulty, including a vast range of behavior: from the repetition of acts that seem to be harmless or even helpful to repetitions that are highly destructive. But all addictions share a feature: compulsion. The person may or may not desire to stop. Frequently, one who desires to stop, could—it seems—do so, but is unwilling to take the necessary steps. Will is disengaged from understanding and desire. A cavity appears within the self. But this means that the unity of body-self and world disintegrates. Our acts no longer gain ecstatic purchase on the world and satisfy primal needs. Addictive repetition leads finally to exhaustion and somnolence. John Muir:

> Once I was let down into a deep well into which choke-damp had settled, and nearly lost my life. The deeper I was immersed in the invisible poison, the less capable I became of willing measures of escape from it.[30]

Real freedom is not some ego-self issuing allegedly rational decisions from its supposedly isolated grandeur; but rather the capacity to decide that springs from rootedness in the enwombing world, from a feeling for what must happen if one is to grow. Only within this evolving musclelike tissue, this felt necessity, is will integrated with appetite, need, and understanding, and really free, that is, effective. Mere willfulness excites rebellion not only in others but within oneself. Willfulness substitutes addictively for the quiet excitement of growth, the mystery of growth, as Black Elk put it.[31] One is not compelled in the sense of moved by what must happen within organism-environment circulation if there is to be a growing, cohesive self—that is, one is not *com*-pelled. One is merely compulsive. A kind of sacrament in which ego is sacrificed has not occurred. The willful will to engage results in disengagement from the world and within oneself. Emily Dickinson:

> *I felt a Cleaving in my Mind—*
> *As if my brain had split—*
> *I tried to match it—Seam by Seam—*
> *But could not make them fit.*[32]

Some of us are not fully in possession of ourselves. Something else is at work. Addictions are forms of possession, some of them demonic. We no more comprehend our state of being than a bull in a china shop comprehends the china shop, and we plunge destructively.

But more understanding (standing-under and bearing) may be possible. We might come to realize that the universe pouring every instant through this node of itself that is ourselves is mainly unknown. It must exceed our comprehension and control, though not necessarily our empathy, intuition, religious and artistic appreciation. As it engendered us it engendered our primal needs; it is right that we have them, right that we fully and ecstatically *be*. Thomas Berry:

> We are indeed into a deep cultural pathology. The ecstatic experience functions as the basis of all civilizations. When the power of ecstasy is subverted into destructive channels, then as in the late Roman world, we are into a disastrous situation, to be cured only by a healing experience at the ecstatic level. There is no way that purely rational processes can provide a remedy.[33]

Notes
1. See S. Kellert and E. O. Wilson, *Biophilia Hypothesis*, 32.
2. Frederick W. Turner, *Beyond Geography*, 9.
3. Hans Selye, *The Stress of Life*, and *Stress Without Distress*. "The name *endorphin* . . . is a contraction of *endogenous* and *morphine*" (Eric Simon, in *Substance Abuse*, ed. Joyce H. Lowinson, et al., 196).
4. "Though we cannot deduce [anglo saxon] *wild* from anglo saxon *willa* . . . will, we can refer them to the same verb to *will*, once a strong verb and of great antiquity . . ," W. W. Skeat, *An Etymological Dictionary of the English Language*, Oxford, 1884.
5. Concerning the excitement of the new, and studies that show how men's and women's hormone levels jump when new, attractive individuals appear, see Diane Ackerman, *A Natural History of the Senses*, 305–307. The John Donne quotation is from "Going to Bed."
6. Personal communication from Buffie Johnson (Mrs. Gerald Sykes).
7. Fyodor Dostoevsky writes from personal experience: "The point is that one turn of the wheel could change everything . . . What am I right now? *Zero*. What can I be tomorrow? Tomorrow I may rise from the dead and begin a new life! . . . [I will dare to risk my last cent and will be] once again, a man among men! . . . Of course I live in a state of continuous excitement. . . . Yet at the same time I have a feeling that I have grown numb, somehow, as though I were buried in some kind of mire"(*The Gambler*, 187, 189). See the contemporary addiction therapist, Rozanne Faulkner, "I failed to recognize the severity of gambling until I visited Las Vegas. Looking into the blank faces of those in the casino at three in the morning told me that gambling drains life" (*Therapeutic Recreation Protocol*, 54).

8. Cited by Alberto Martini, *The Great Artists, Bk. 1, Van Gogh*, New York, 1977, 1.

9. At what point stages of awareness arise is speculative. Yet brain science has established that the fundamental levels or components of the brain (popularly: "reptilian, mammalian, cerebral") were laid down in the phyla in that order over millions of years, and the development of the individual (ontogeny) roughly recapitulates the development of the phyla (phylogeny). That we inherit tendencies from the dim past is no longer doubted by science. But there is no unified research program to pin them down.

10. Fast-moving research on the sleeping-waking cycle reveals that "Mom is functioning as the transducer for the fetal circadian system. She takes in light information to her circadian system, and then that is communicated to the fetal circadian system . . . Melatonin is a small molecule, and it slips readily across . . . the placenta"—from "Awakening to Sleep" (*New York Times Magazine*, referred to in the Prologue, 41). More generally, circadian rhythm genes have been discovered in fruit flies and bread mold. In mammals the primary pacemaker is in the brain, but there appear to be "regional clocks" in tissues outside the brain. *Science News* 151 (May 17, 1997).

11. See my "Mimetic Engulfment and Self-Deception."

12. See Alexander Marshack, *The Roots of Civilization*, on Paleolithic practices of time factoring.

13. *The Relevance of Philosophy to Life*, 109.

14. *The Maine Woods*, "Katahdin." This has been republished with helpful notes by William Howarth, *Thoreau in the Mountains*, New York, 1982, 150.

15. Igor I. Sikorsky, *AA's Godparents* and Terence Gorski, *Understanding the Twelve Steps*.

16. "Building a Railroad Deep Into Brazil," *New York Times*, January 19, 1995: "Sao Paulo . . . High in the corner office of glass sheathed office tower here, Olacyr de Moraes passed a ballpoint pen across a map of Brazil, tracing a new railroad network westward through agricultural lands, then north across the Amazon. . . . 'My mission is to build this railroad' . . . he once worked for 20 years without a vacation."

17. Peter Nabokov, *Native American Testimony*, 184.

18. See Anne Wilson Schaef, *When Society Becomes an Addict*, and Dolores La-Chapelle, *Sacred Land, Sacred Sex*.

19. *Dr. Atkins' New Diet Revolution*, New York, 1995. For a somewhat different approach to avoiding "bad carbohydrates" and balancing hormone and insulin levels, see Barry Sears, *Enter the Zone*, New York, 1995. Also Rachel and Richard Heller, their excellent *The Carbohyrate Addicts Diet*, New York, 1991.

20. Catherine Sneed, head of the San Francisco Garden Project, the Schumacher lecture at Yale, October 21, 1995. I am indebted here to David Ehrenfeld.

21. For sheer quantities of things consumed, see Deborah Winter, *Environmental Psychology*, ". . . *overconsumption is the biggest depleter of the earth's carrying capacity.* Americans consume, either directly or indirectly, over 100 pounds of raw materials a day, including 40 pounds of petroleum, 25 of agricul-

tural products, and 20 of forest products. This voracious depletion occurs because in the past century we have transformed our households from places that produce necessary objects to places that consume convenience objects" (21).

22. In a talk, "Higher Education for a Just and Sustainable Future," sponsored by Center for Respect of Life and Environment, Washington D.C., September 15, 1995.

23. See Eugene Gendlin, "Thinking Beyond Patterns."

24. Also note Ilya Prigogine, "Mind and Matter: Beyond the Cartesian Dualism," in Karl Pribram, *Origins: Brain and Self-Organization*. Pribram comments on Prigogine's idea of possibilities: "[He demonstrates that] temporarily stable orders can be formed out of apparent chaos. These stabilities far from equilibrium are the stuff life is made of" (ii). Body-self thrives on the not fully predictable to maintain its equilibrium regeneratively.

25. Emerson in Ziff, *Emerson: Selected Essays*, 228. Unless otherwise noted, all references to Emerson are to this edition.

26. See J. Krishnamurti, *Freedom from the Known*. For exposure of "natural" cartesian subjectivism, see J. Baird Callicott, "Rolston on Intrinsic Value: A Deconstruction," *Environmental Ethics* 14 (Summer 1992).

27. See Herman E. Daly and John B. Cobb, Jr., *For the Common Good*, 87–88, 91–92.

28. See Dolores LaChapelle, *Sacred Land, Sacred Sex*, her many allusions to the International Growth Society. See also Daly and Cobb, *For the Common Good*.

29. See the televised interviews with Joseph Campbell done by Bill Moyers, and published by the latter as *The Power of Myth*, New York, 1988, 106.

30. *John of the Mountains*, 192.

31. See John G. Neihardt, *Black Elk Speaks*, 208.

32. Quoted in Susan Griffin, *Woman and Nature*, 94.

33. Personal communication.

Chapter Two

Rediscovering Space,

Time, Body, Self

But through the phases of perturbation and conflict, there abides the deep-seated memory of an underlying harmony, the sense of which haunts life like . . . being founded on a rock.

—JOHN DEWEY, *ART AS EXPERIENCE*

I was sitting at my desk in the university one bright afternoon in 1970 with a stack of work half finished before me. Time was short. The phone rang and the excited voice of our five-year-old daughter sounded unexpectedly. She and my wife had discovered a dog running loose in the library; they had grabbed the fragment of rope hanging from its neck, and now the dog—skin and bones—had to be fed and had to have a home. Could she please, please keep it?

"But who will take care of the dog?"

"I will, I will!"

This was not entirely implausible, since she loved animals and was fairly self-disciplined. I noticed someone standing in the hallway awaiting me, as well as the stack of unfinished work on my desk.

"All right, you can keep it if there are no tags on it and nobody claims it."

"Oh, Daddy, you won't be sorry! We'll give him a bath right away!"

She hung up, and it was only later that night that I saw the dog himself, now dubbed "Buppie." He stood in the dim garage. He was huge—she had not spoken of the size—and though recently fed and with most of the filth cleaned off, had a very woebegone—but friendly—look. He was perhaps three-fourths German shepherd, the remaining fourth indecipherable: his ears, furrier than a shepherd's, did not stand up straight.

We built a pen for the dog in the back yard, and for awhile our daughter and our nine-year-old son fed and cared for him. But Buppie was so powerful that they could not enter the pen without him knocking them over and licking them. The fact was, only I could half-way control the dog. In my days at home I would look out the window and see his face peering out lonesomely from his pen, so I got in the practice of running him through the neighboring woods.

Actually, he ran me. When a long limber chain was attached to his collar and he was let out of the pen, he would dash so powerfully into the woods that I had to run as fast as I could to avoid falling. After several minutes of frantic running, Buppie would mill around in the woods, sniffing trails that were evident to him, not to me. After several minutes I could sit down without the dog straining madly at the leash. What I could smell—the woods—became evident, and Hopkins' line came back: "There lives the dearest freshness deep down things."

Panting deeply, sitting against a tree trunk, I was unkinked, rejuvenated. It was fairly quiet in this little woods, and I could hear but dimly an endless roaring in the background. Only late at night was the roaring of traffic on the New Jersey Turnpike, barely two miles away, grossly perceptible as the constant background of my life—trancelike, relentlessly repetitious, anonymous motion.

One day the dog led me across the street and down a lane at the foot of which stood a small water-pumping station. It was surrounded by dense undergrowth, and the dog pulled me into it. Avoiding thorn bushes, we picked our way. Following Buppie, I stepped gingerly through a final densely branched bush and stood stock still. So did the dog.

There before us was a sight that strained belief. It was if we had stepped through a time-warp and stood instantly two hundred years earlier and a thousand miles away on a frontier. In a vast arc from left to right, no human habitation could be seen, only farmers' fields cut into the woods, forming a pattern that stair-stepped away into the distance. I had often traveled the road that lay about forty yards to my left, paralleling this area, but had never looked through the dense but narrow growth that bordered it. Such a vantage point, just thirty miles from Manhattan, had been unimaginable. My gaze now swept the horizon for about 150 degrees, stopped only by the fine and misty lines of trees on the farthest of the hills that undulated away into the distance.

There was strange dilation and release. The horizon became evident as hori-

zon, that is, it became clear that the horizon of the farthest hills enclosed all that could be directly seen, but beyond that lay everything else. Although, what all was there, and just what sorts of things might be there, I had no idea. We belonged somehow to a fathomless whole, to the universe.

Dreamily the dog led me on through these beckoning fields and woods.

Genesis, chapter one, verse twenty-eight:

And God blessed them, and God said unto them, Be fruitful, and multiply, and replenish the earth, and subdue it: and have dominion over the fish of the sea, and over the fowl of the air, and over every living thing that moveth upon the earth.

The Earth, after all, is a difficult place. To live we must wrest our sustenance from it. It seems self-evident, then, that we must subdue the Earth or be subdued by it. Equally self-evident, it seems, we are the most intelligent and valuable of Earth's creatures; from which follows the ultimate mandate, that we should have dominion over all other living things.

The basic confidence that dominion is justified, whether divinely ordained or not, is still in force.[1] If overpopulation is a problem it must be identified, solved, conquered. When the Hebrew tradition intertwined with the Greek and emerged through Rome as Christianity, the attitude of detachment from Nature and domination over it fed the rise of modern science. The obvious success of an experimental method that could make precise predictions, test for their fulfillment, and build machines to bring about results on command fed on itself.

Who could doubt that knowledge of Nature is valuable? Or that buildings against the winter's blast are valuable, or weapons against marauding beasts, engines to lift back-breaking loads, communication devices, etc.?

Yet we now know that prediction and control for short-term gain has, say, altered the atmosphere of the planet. Millions of spray cans firing in the air (the gathered weight of individual banalities—hair spray, dog spray, scent spray—is tremendous) corrupt the ozone layer that protects from lethal solar rays; discarded air-conditioners and refrigerators, spilling their gasses in junk heaps, do the same. Millions of automobiles and trucks burning fossil fuels contaminate the air so heavily that the heat of Earth is trapped, posing the probability of polar icecaps melting, seas rising, coastal cities being destroyed, loamy soil drying out, the national supply of food dwindling. Moreover, the

Greenhouse Effect is changing the timing of the seasons, delaying their onset in some locales and bringing about an early spring in others.[2] Since basic cultural functions, not just farming, are seasonal, these changes bode to be disruptive on a broad scale.

At issue is an attitude of domination that can't begin to be questioned because it is all-pervasive, constant, contrastless—a kind of trance. Our ability to control Nature and to produce "goods" changes Nature so irruptively that standards for evaluating what we are doing are outrun.

"As right as rain." But since the rain in many sections is so acidified from burning fossil fuels that it kills fish and trees, what is right about it? "Practice makes perfect." But if steroids injected illegally in athletes allows them to outperform others who practice diligently but don't take them, what becomes of our maxim? "As frightened as a mouse." But if gene splicing recreates them as voracious ratlike beasts, what then? "As nourishing of her young as a cow." But if fetal implants engender offspring so foreign to the mother that she lacks all affinity for them, what then? "If winter come, can spring be far behind?" But if the predicted Greenhouse Effect does transpire, there will be little winter to speak of, thus little fallow time, and diminished promise of rebirth in the spring.[3] A psychoactive pill may eliminate grief over the death of a loved one. But what if this eliminates grieving, the closure that opens the way to a new start in life? Inability to evaluate means inability to act effectively, inability to *be* fully.

Our technological means of control may have gone permanently out of control, the attempted subduing of Earth subduing us. If this is true, disorientation and rigidity have set in (even in the midst of frenetic activity), for we are propelled by an attitude that has destroyed its own guidance system.

We need to know how to evaluate ourselves. If we have no idea of where we're going and why, the temptation to grab mechanically for immediate gratifications is uncannily strong. The beer commercial says it: "Grab for the gusto. You only go around once." Believing there is only one burst of glory eclipses the possibility of many cycles in a lifetime, each waning one followed regeneratively by another in the fullness of time—until death. Nature cannot be trusted, it is thought, but must be forced to climax at every turn. This is addictive behavior.

Nature a Machine?

Industrial-technological life, its production and marketing of commodities entwining us routinely, began with the great seventeenth-century thinkers who discovered the precise principles of machines, "mechanics." Sir Isaac Newton wrote, *hypotheses non fingo* ("I do not cast hypotheses"—as one would throw dice). He meant that as a physicist his hypotheses are not arbitrary but emerge from a groundplan in which the only possibilities projected for Nature are mechanical ones. The rich meaning that things have in our prescientific, spontaneous bodily experience is masked out. Particularly, our experience of "the everything else" is not measurable, so the science of Nature must overlook the gut sense of the World-whole and our role in it.

The measurable forces of Nature, such as gravitation, were determined by configurations of masses of matter; these then determined their own changes, so that the circle of scientific thought was closed. This mechanistic theory of Nature was plotted within a space and time construed abstractly, on linear metrics and coordinates alone. Closed off was the circumambient world of sensuous presences and, of course, the spontaneous feelings of belonging to it organically—feelings of renewable vitality, centeredness, confidence.

Admittedly, there had to be an observer to mark positions and velocities of masses and to make measurements, but the abstractions within which mechanistic physicists work do not require them to account for their own perceiving or measuring. As long as observations through instruments can be repeated by others, differences in individuals' consciousness can be cancelled out. Indeed, in Newton's view, consciousness itself can be half-way forgotten. Space, denuded of all "subjective" experience, is merely an indefinitely large, precisely metered box, possibly entirely empty, but in fact holding within it inert matter in various positions and kinetic states. The box rolls along a precisely metered linear track of time inexorably. Not only can consciousness be ignored, but, of course, self as well.

The gains from this attitude of abstractness and detachment are immense: much contemporary technology stems from seventeenth-century physics. Through the application of universal principles of motion, matter, and force, the Earth is reshaped, and vast goods-for-sale, commodities—and the wealth for some to buy them—are produced.

But the price paid for these gains is at least equally great though much less obvious. For when emotional involvements are reduced to supernumerary, subjective status, the foundation for values collapses. Goodness satisfies the human heart, regenerating us ecstatically within the unplumbable whole and the long haul. This is a fact. Emotional involvement as a whole self within Earth, sky, and horizon is self-evidently,

intrinsically valuable. But mechanistic physics lacks the concepts for dealing with such facts. Force, momentum, acceleration, geometrical shape
are not sufficient. Our bodies are not just mechanisms to be manipulated.

Ignoring the foundation of all value, physics must focus only on the
values it can deal with: efficiency of means for achieving ends—
instrumental, not intrinsic, values. The ends themselves must, within
this framework, seem arbitrary. Or, since this is not a pretty admission,
thought dominated by physics and technology tends to assume obvious,
more or less measurable, if crass, ends-values: pleasurable bodily sensations, degrees of comfort, human longevity, buying power, etc. But
when meaning shrinks and life's energies go to ends less than intrinsically satisfying and enduringly ecstatic, we exploit and fracture ourselves. We feel vaguely guilty.

Technology prescribes instrumental values and only the crassest
ends values. If we want B, means A can be used to get it. But if we
cannot imagine ends intrinsically satisfying, technology has nothing at
which to aim and wanders and plunges. What has it availed us if "we
gain the whole world" but lose even our desire for sacramental experiences in which ego is sacrificed, leaving room for the pulsing and cycling presence within us of the World itself—the homecoming
experience of ecstatic belonging? Technology, our means of control,
goes out of control.[4]

Technology: Power and Alienation
Seventeenth-century physics and technology birthed the modern world
abruptly. Systems of control are created that have a life of their own,
wildernesses with their own will. But they are utterly unique, for they
have been created by ourselves, and very recently. They defy Nature's
wildernesses in which we evolved and took shape over millions of years.

We haven't learned to adapt to our own adaptations and creations,
to preserve age-old intrinsic and unconditional values along with the
new instrumental ones. When instrumental genius supplies gratifications that substitute for primal gratifications of primal needs, we become entrapped. When, for example, our need for restful sleep with its
rapid eye movements and dreams is overridden by quick fixes, this is
addictive adaptation, dependency.

We can't yet combine technological control with skillful and awed
abandonment in regeneratively cycling Nature. For the only uncondi-

tional and intrinsic values are the satisfactions of growth as whole selves in the whole world.

René Descartes, Modern Science, Human Difficulties

Descartes' version of seventeenth-century physics and technology is perversely ecstatic, and contributes to the dilemmas of the modern world. We are supposed to be hybrid beings: godlike minds or consciousnesses that hook up somehow to mechanical bodies. We stand off from our bodies in "transcending" calculative thought about all bodies, predict their behavior and manipulate them technologically and industrially—irruptive ecstasy. Other animals are only mechanical bodies. As a machine may screech and scream and feel no pain, so an animal.

In one stroke humans are dislocated from their own bodies and from the bodies of all nonhuman creatures as these have evolved together with us in regenerative Earth in the most intimate communion over countless millennia. The temptation to treat our bodies as mere machines for generating pleasurable sensations is strong. But if selves are body-selves, using the body as a mere means is violating its intrinsic value: it is violating ourselves.

Our bodies—body-selves—bring needs for intrinsic value, needs to grow ecstatically and develop. These are "programmed" in stages pretty well determined genetically through the development of the species *Homo sapiens* over approximately a hundred thousand years (and when we include the genus *Homo*, much longer than that). To be true to our sources is to be true to ourselves—this the root of the sacramental and sacred.

We are organisms that grow to ripeness and maturity if properly respected and fostered. Mechanisms simply wear out. Our capacities need to develop, and they were ours long before modern technologies that confined our roles. The danger is to suppress primal needs, to shoot wildly, addictedly, trying to satisfy ourselves.

The development of body and the development of mind go hand in hand—pun intended. If nurturing fails at a crucial point in the stages of development, the window of opportunity closes. When infants are placed in impoverished environments, their abilities to explore, observe, and manipulate materials at crucial stages are blocked. Hence their ability to form and manipulate concepts is blocked also. Jean Piaget writes of the operation of convertibility essential to the concept of the identity of things—of their reality. For instance, when the child rolls a piece of clay into various shapes, and then rolls it back into earlier ones, the idea arises in time that the volume of clay—the clay itself—remains the same despite the different shapes given it.[5]

There have been a few cases of feral children, whose crudest needs are satisfied somehow, but who are isolated from personal contact with humans. Because the window of opportunity for learning a human language closes without their learning one, their abilities to learn are radically curtailed. And since they do not learn to stand, their buttock muscles do not develop on schedule, and possibilities for human life that are related to the upright posture are foreclosed or greatly diminished. In the case of nonhuman animals, scientists feel free to experiment. Joseph Pearce writes:

> If a kitten spends his first few critical weeks of development in a compartment with vertically striped walls, the grown cat will be able to see only those objects of a vertical nature. He will avoid the legs of a chair perfectly well but will run smack into the horizontal rungs.[6]

Pearce notes that we are not kittens. Yet optimal development of the human individual can occur only within successive favorable matrices:

> *Matrix* is the Latin word for womb. From that word, we get the words *matter, material, mater, mother,* and so on. . . . The womb offers three things to a newly forming life: a source of possibility, a source of energy to explore that possibility, and a safe place within which that exploration can take place. Whenever these three needs are met, we have a matrix. And the growth of intelligence takes place by utilizing the energy given to explore the possibilities given while standing in the safe space given by the matrix.[7]

It is only because of the mature adult's ultimate matrix—the reliable and stable if demanding Mother Earth—that the skill-endowed parent can supply a reliable matrix for the child.

Mother and "Mother" as Matrices
In initial stages of maturation we are tied into the Whole via the mother.[8] The first stage is fetal, the second the nursing infant. Through the support, energy, and stimulation she offers, first as womb and then as encompassing warm body, infants are launched into those playful exploratory activities that lead to a sound sense of our individual being within an interdependent world, finally to our maturity, our ecstatic responsibility. Successful initiation and support is a primal need at each stage of maturation—for example, the prepubescent latency period

from about seven through eleven, in which the child experiments and explores with spontaneity and abandon, and it is safe to do so. Given success at this stage, and at the equally crucial next one of adolescence, the visceral sense of the mother as matrix gets transformed and broadened to become the sustaining "Mother Earth": the world experienced as there and real whether we are around or not, the world that will sustain us if we are sufficiently skilled and respectful (and fortunate).

Adolescents mature when they move from mother to Mother Earth, from one they can blame to one that as mature adults they cannot. Julia Kristeva delineates how essential is the mother's confident relatedness to the world. If she focuses exclusively on the child, she will be abject, and so will the child.[9]

Earlier I wrote of male seduction of women as adventure in will-of-the-place and the unknown, and of the possibility that seducing will become a way of life that closes in upon itself and degenerates into addiction. Ecstatic adventure can perversely miss the mark. If the male has prematurely separated from the mother, the greatest of all primal needs has not been satisfied. An initiation that roots him in the world via the mother has not been performed. The need cannot be clearly identified, for the identification of anything depends initially on the mother's (or motherly person's) attentive identification of it. But opaque and unthematized, this need persists, thrusting blindly for satisfaction.

And also fearfully and angrily, for when the one source that might satisfy a hunger fails for some reason to do so, this must arouse fear and blind rage in the hungry child, and these feelings remain. "The child is father of the man," wrote William Wordsworth.

Pearce sees the connection between rape and the primal, wild, but frustrated urge to bond with the mother:

> Sexual hunger has little to do with why men rape. Nor does sexual attractiveness in its usual social sense play a role; the eighty-year-old woman is just as liable to be raped as the twenty-year-old. The rapist himself does not understand the hunger that drives him.[10]

If the child does not bond with his or her mother, it is likely he or she will not bond with Mother Earth. If this pattern of nonbonding becomes a traditional communal attitude, as it tends to do in technological and patriarchal North Atlantic culture, the generalized rape of Nature follows.

Erik Erikson, consonant with Pearce, and referring to the work of Joan Erikson, writes of the founding level of human identity:

> The newborn, recumbent, is gradually looking up and searching the inclined and responsive face of the motherly person. Psychopathology teaches that this developing eye-to-eye relationship is a "dialogue" as essential for the psychic develpment and, indeed, survival of the whole human being as is the mouth-to-breast one for its sustenance. . . . [A] lack of such early connection can, in extreme cases, reveal an "autism" on the part of the child that . . . probably is responded to by some maternal withdrawal. If so, we can sometimes observe a fruitless exchange, a kind of private ritualism characterized by a lack of eye contact and facial responsiveness and, in the child, an endless and hopeless repetition of stereotyped gestures.[11]

In this failure to meet a primal need for recognition and confirmation lies a deep root of addiction. The mother-child greeting ritual is the Mother of all subsequent rituals that initiate individuals into the successive stages of life. If it never jells, Mother Earth will not be greeted respectfully, and, of course, the needs evolved in Nature and capitally one's own will not be felt to be such. They will emerge in substitute or counterfeit form, with an element of "not me" about them. Trance-like repetitions occur and addictions flourish.

Ecstasy Deprivation and Alienation

The cruel, semihidden irony of technological "progress" is that the very ease with which we push buttons and turn dials—or recruit auxiliary care-givers—makes "unnecessary" the development of the full array of human feelings, skills, and satisfactions of primal needs. Or easy access to specialists' knowledge may make "unnecessary" much body-self learning that is development toward coherence and maturity.

Disconnection from Nature in the name, ironically, of natural science leaves a bottomless pit. Addictive behaviors pour into it. We see its effects in the wasting of Nature and the children of Nature, indigenous peoples and their profound contentments.

In 1938 Carl Sauer noted how the ideology of technological productiveness, its "realism," had become a monomaniacal ecstasy that threw us out of touch with our own reality as sensuous beings capable of multidimensional ecstatic lives of gratitude and awe. Inheritors of North Atlantic civilization cannot tell the difference between yield and loot, says Sauer.[12] Black Elk, native American holy man and thinker, spoke of

Europeans' reduction of Nature's gifts to commodities—buffalo hides, buffalo tongues, timber, and so on. Gold was "the metal that drove them crazy."[13] Ultimately Europeans, for all their apparent successes, reduced and alienated themselves; they mangled their roots in Nature.

Isabel Kent of the Ute Tribe writes of the traditional prohibition of selfishness:

> If you've seen a chokecherry tree . . . The upper part is for the birds. The middle part is ours, the human's, the rest, the bottom part, is for the animals that cannot climb. That's why we were taught that we don't pick the whole thing.[14]

Marcus Aurelius, Stoic emperor of Rome, put it this way: "All things are implicated with one another, and the bond is holy."[15]

It seemed self-evidently true to seventeenth- and eighteenth-century scientific thinkers—as well as to most of their descendants in North Atlantic culture today—that it was good to debunk practices such as winter-solstice rituals that aimed to halt the sun as it sank deeper in the sky through autumn months. By refusing to perform the rituals, it could be seen that the sun would not continue to sink in the sky day by day, but after a certain point would reverse itself and rise back toward the zenith, toward spring, summer, and the reawakening of life with no effort on people's part. The calm and detached view of science proved that human involvement with the sun was inessential for its behavior.

But the price paid for this truth was the obscuration of another broader, older, and deeper one: that ritualized participation in the regenerative cycles of Nature is itself regenerative.[16] In their seasonal celebrations—at the shortest and the longest days of the year, for example—indigenous peoples believed they were participating in, and contributing to, the renewal of all things. The movements of the heavens and the cycling seasons were not just "out there," but unfolded excitingly within their own bodies as well, and out of this vast fertile ground sprang times of planting and harvesting, hunting and tanning, fashioning tools and using them, and so on.

Without an all-inclusive ritual matrix in which human development is keyed to participation in Nature's cycles, things tend to fall apart. There is no way of sorting out ecstatic involvements into wild, constructive, regenerative, on the one hand, and wild, disintegrative, degenerative (merely shooting wildly), on the other.

Belonging to Nature, Belonging to Self

The primal fact at all times and places is reciprocity. We are nourished only to the extent that we nourish (the responsive infant nourishes the attentive motherly figure). Streams sing to us and lead us on, and a parent can guide a child through time-honored ways of fishing, say, and of returning home with the catch, and the child nourishes the parent. Mountains beckon, bolster, or block us; billowing silver clouds buoy us; trees shelter and comfort us—or threaten to fall on us. In this reciprocal vibrancy the body self's stances and feelings are tied in the most intricate way to its surround, and how we expect it to circulate and feed back into us rhythmically.

Dewey alluded to "deep-seated memory of underlying harmony," which he likened to "being founded on a rock." The memory may now be too weak to help much. David Orr writes that "something pulls at us, but weakly."[17] Edward Casey writes of the

> insidious nomadism endemic to modern times in which the individual, subject to disorientation and *anomie*, drifts within the desolate and indifferent space of housing developments and shopping centers—the architectural heirs to the gridded, isotropic space first posited by Descartes and Newton.[18]

With all our scientific guidance systems, technological marvels of transportation, communication, action at a distance, we may have lost touch permanently with our regenerative base in Nature. Unable to articulate this possibility, and confined to technological and scientific horizons, many seem to imagine only one solution: to objectify the body and focus on the brain, in isolation of the regenerative Nature in which we adapted habitually and took our structure over millions of years. That is, to intervene directly in the brain with schedules of psychoactive chemicals that ignore traditional rhythms of sustaining interaction with the Earth and their attendant stories and rituals.

But it is through these that the framework of evaluation that shaped human life was formed. There is much evidence now that scientific and technological interventions in Nature's regenerativity are undermining it. If the ultimate Matrix fails, everything else fails with it.

Scientism

Twentieth-century science has made tremendous discoveries, from intergalactic structures to electrochemical features of the brain. But, caught up in its own ecstasies, science has become an ideology that defines ever-spreading North Atlantic influence—a quasi-religion. Science cannot prove that it alone can reliably formulate meaning and know—for example, know what's right to do, or what the lived quality of a situation is. The belief that it can is scientism.[19] To prove it science would have to show, first, that all meaningful issues or questions can be formulated by itself. And, second, that only science gives reliable answers.

But even to claim that science knows all meaningful questions begs the question of its ability to create meaning and to know. Facing the question responsibly would require science to enter into nonscientific modes of inquiry—artistic ones, for example—and to show they get us nowhere. But to do so it would have to exceed its methods and evidential base.

Limiting its idea of world to what it can calculate, much Western science and technology is still fixated on the seventeenth-century. It denigrates the gut sense that our urgings and actions feed out into a world that has been, is, and will be feeding back into us in ways that are beyond our abilities to calculate. It denigrates the womblike wil-derness world, mainly unknown, within and beyond our horizons—that sustenance and circumpressure resounding within our bodies. In a culture dominated by scientism, the body-self's complement of needs cannot be met. Our style of life must be addictive.

As now constituted, North Atlantic culture has no story of its origin in Nature, and rampages across the planet. We come, as it were, from outer space, and see Earth as a prize to be possessed, manipulated, and used for our short-term gain and pleasure.

The ultimate irony, of course: to treat one's body as merely a mechanism to manipulate destroys the spontaneity and coherence of body-self—produces the proto-addictive state. We do not necessarily become addicts in this atmosphere of scientism, but effectively countering the tendency requires real freedom—the ability to live fully and creatively. The dangers of mere libertinism and anarchy are great.

Scientism/Science

Scientism of course is one thing, science another. The role of various twentieth-century sorts of science in disclosing the nature of health and maturation will be discussed in the second section of this book.

For now it is the role of scientism that concerns me. Consider how some parents are sometimes reluctant to judge—or even inquire into—their childrens' behavior, for they say, "I'm not a psychologist." Or, again, many bow before supposedly scientific economists and refuse to consider that they dislocate us from regenerative Nature and our own organisms. Mechanistic physics left over from the seventeenth-

century—and ever new modes of technology and commerce—expand
our powers in obvious ways, and shrink them in furtive ones: strange
possessions that subtly dispossess us.

*The dog pulled through the undergrowth and the field opened before us. The
field! We ran down the rutted path that counted as a road for farm machinery
and passed the roundish trees that once surrounded a house. We often poked
among debris here and bits of foundation, most of it now covered with soil. I
went my way, and he, at the end of his supple chain, went his. My feet exposed
a bent and shattered pair of "granny glasses," the real things—worn before they
had become chic—their fragmented lenses turned faint violet from the sun's rays.
On a trip the previous spring a corroded 1895 nickel turned up. But neither
the dog nor I found anything more this time, and he strained on the chain down
the road.*

*As we walked, a pheasant erupted noisily from the bracken bordering the
fields and startled us. Lifting a body that seemed too heavy to fly, it propelled
itself like a rattling shot across the fields.*

*The dog slowed a bit and we walked around several of the fields on the rutted
roadways, then passed through a small opening in the thorn bushes and into a
dense square of forest several hundred yards on a side. It was a gloaming space,
vines festooning the trees, and many years' accumulations of leaves and deadfall
lying on the ground. The Earth was digesting them. (Once in the woods next to
our house I encountered a neighbor. Chasing our respective sons for creating
some mischief, we had been thrown together for the first time. Generations of
trees grouped around us, standing and fallen, the dead feeding the young. He
said that the woods would be better if all the dead branches and sticks were
cleaned up.)*

*The dog's awareness was, in a real sense, transferred to me. We became an
excited, moving part of the place we experienced, absorbed in the immediate
surround, entranced, although I had not lost the capacity for self-consciousness.
I might have noted that I changed my position with respect to things that themselves did not change. I might have experienced my body focally as my own, as
I-myself, thereby experiencing myself as separate or distinct from the rest of the
world. (This capacity Buppie did not share, presumably, or possessed only in
nonverbal ways.) But I refused to exercise the option!*

*The dog and I were bound tightly to each other and to the surround. The
world buoyed us in its life and propelled us in its power. Every sound and sight
from every direction fixed us and was registered; each possessed us delightedly.
We picked our way along, the leaves and twigs sounding too loudly beneath us,
and found our way to a favorite spot, about forty feet in diameter. I called it
"the clearing."*

By this time the dog lay quietly, panting easily, his restlessness dissipated. I sat against a fallen, moss-covered tree and looked into his friendly face. After some time I became immobile and quiet, the inner chattering monologue emptied out. Only then was I aware that it had been going on constantly. Thinking of nothing in particular, I lost my customary focus. I was empty, immobile, yet the surrounding world was vibrantly present on the fringes of consciousness. It flowed in with its larger life, exciting, intensely alive, and reassuring.

An indeterminate time passed. It must have been fairly late in the afternoon, for the woods were filled with dusky twilight. Suddenly I was alerted to something peculiar in the field of vision. About a half-dozen blurs of white moved in gently swooping, scalloped sequence above the forest floor and behind some trees roughly two hundred feet away. I raced to find explanations for what it was that moved. The best was this: a sequence of white birds had flown through the woods. But I had never seen such birds before. I could only think of movies of cockatoos in Australia, and they wouldn't be those.

It was not until one day six months later that what they were became clear. Buppie and I had traveled to the farthest point of our range. Climbing a hill we posted ourselves at the top amongst scattered trees and large, slatelike stones. This was "the watch tower." From this point we could look out at housing developments in the distance which over a decade would encroach upon and finally consume the fields and woods we traveled. Again mind emptied out; I became quiet and immobile yet filled with intensest life.

It was as if I were going back to the sources of life. Had ancient ancestors, encoded in my body, been touched again into being? We flowed with the majestic course of the day.

The afternoon wore on, and since we were some distance from home, I finally decided to start back. But wanting to go a new way, I started down a side of the hill not traveled before. We picked our way through an unexpectedly extended area of thorn bushes and stood panting.

On the right fringes of the visual field an incredible sight unfolded. A line of small deer, about a half-dozen in single file, pranced delicately along, coming close to us, about thirty feet away and looking at us, and just as quickly passing by. They had large white tails that moved in scalloped sequence through the woods.

These I named "the ghostly herd," and told the dog I had done so. We would repeatedly seek them as the years went by.

Notes

1. There is a difference, of course, between domination understood within the context of divine ordination, and that which falls outside it. In the former,

it is understood within a context of other divinely ordained acts or duties, such as piety, restraint, stewardship.

2. "You'd have to be crazy not to notice it" [the shift in the seasons]: David Thomson, AT&T researcher who developed a new way to statistically analyze data collected over many years. His findings to be published in *Science* (*Newark Star Ledger*, February 21, 1995).

3. A few scientists disagree about global warming, for example, Arthur and Zachary Robinson in *The Wall Street Journal*, December 4, 1997, "Science Has Spoken: Global Warming Is a Myth." It is tempting to dismiss this as a case of "pay your money and buy your expert witness." But the real defects of their argument can be seen within the argument itself. They point out that solar activity varies considerably from decade to decade, even year to year, but it has trended upward since 1750. They observe that in the last twenty years atmospheric temperature has declined while greenhouse gases have increased. This they regard as the clincher. But according to their own statistics, twenty years may be a mere blip in temperature changes. Within this time atmospheric temperature might have declined more if greenhouse gases had not increased. The fact is, there is no way of telling with certainty what our massive interventions in Nature have done. It would seem to be wise, however, to err on the side of prudence. The Robinsons might embrace some scientific and scholarly caution before they assert that increased greenhouse gases will promote plant growth and are therefore good for the planet.

4. "No single man, no group of men, no commission of prominent statesmen, scientists, and technicians . . . can brake or direct the progress of history in the atomic age. . . . The approaching tide of technological revolution . . . could so captivate, bewitch . . . that calculative thinking may someday come to be accepted . . . as *the only way* of thinking." Martin Heidegger, *Discourse on Thinking*, 52, 56.

5. Piaget and Inhelder, *The Growth of Logical Thinking*, xiii, xx, and indexed under "reversibility."

6. *Magical Child*, 6.

7. *Magical Child*, 16.

8. Or via whoever plays the motherly role. After birth, other persons can occupy this. I do not always make this point when I speak of "mother," but it should be understood. Chapter eleven maintains that whoever this is is caught up in the force of the archetype, Mother. Are women, on average, better suited to this than men, on average, at this time? Probably, but the issue is complicated. For "archetype," see chapter 9.

9. *The Kristeva Reader*, "Freud and Love," 241 ff. To be abject is to be empty. It's not so much that repression is at work, as that emptiness is concealed through addictive talk and other activities.

10. *Magical Child*, 220.

11. Erikson, *The Life Cycle Completed*, 40, 45, and Joan Erikson, "Eye to Eye."

12. Quoted in LaChapelle, *Sacred Land, Sacred Sex*, 36.

13. Neihardt, *Black Elk Speaks*, 8.

14. Jan Pettit, *The Utes*, 28.

15. Quoted in David Griffin, *Sacred Interconnections*, epigraph.
16. On the topic of arrogance, see David Ehrenfeld, *The Arrogance of Humanism*. I develop the concept of body-self in chapter 4.
17. "Love It or Lose It," in Kellert and Wilson, *The Biophilia Hypothesis*, 415 ff.
18. From his manuscript of 1991, *Getting Back into Place*, 770.
19. Concerning the limits of science, Wolfgang Pauli, *Writings on Physics and Philosophy*, 128, 259.

Chapter Three

Circular Power Returning into Itself

❦

The first in time and the first in importance of the influences upon the mind is that of nature . . . There is never a beginning, there is never an end, to the inexplicable continuity of this web of God, but always circular power returning into itself.

—RALPH WALDO EMERSON, "THE AMERICAN SCHOLAR"

❦

I was hunting—probably fourteen years old—threading through creosote bushes, joshua trees, rocks, cacti on the sandy floor of the Mojave Desert. Several days' efforts had brought no luck. But there it was: a coyote loping silently through rocks and bushes. It must have seen or smelled me. I raised the rifle to find it in the telescopic sight. There it was, magnified, a wild, jiggling blur. Finger rigid, body shaking. However hard it was to aim, I had to get off a shot. I fired. It kept running without breaking stride. A desperation shot followed as the animal, slipping through bushes and rocks, vanished.

Chagrin and remorse hit me. Why hadn't I done better in the moment of testing? The incident would not let go for weeks. What impelled me to try to kill the animal? If it had lain there, eyes glazed and fixed, fur matted with blood, I would have felt bad too. Its death wasn't needed for food or for protecting live-stock. What kind of creature was I? What did I want or need?

The nineteenth-century "romantic" reaction against seventeenth-century mechanistic science and ideology was—and is—one of the great confrontations of history. As if the now nearly universal North Atlantic world were a huge container of liquid that rocks in crashing waves, we exist in the turbulence. Everyone knows of tensions between environmentalists and industrialists, between "greens" and those advocating unlimited growth of gross national product, and so on.

Toward the middle of the last century, Emerson epitomized the romantic rebellion. The roots of *romantic* (*roman*) mean the novel or the new as well as renewal. In Emerson and his predecessors these spring up in a burgeoning, blooming, defiant growth. Thinkers and artists declare that modern technology and industrialization have pulled us out of touch with our deepest interests, needs, excitements. If we would find ourselves in the present and claim a future that is our birthright as organisms, we must go back—reconnect with our original sources in the Earth. This sounds as perverse to many in our day as it sounded to many in Emerson's.

Repudiating Mechanism

Emerson reasons that exploiting Nature as if it were a machine means exploiting ourselves. He replaces the metaphor of machine by that of organism. In a machine the parts connect at their hard surfaces to contiguous parts. To understand the movement of these it is sufficient to understand impulses of force communicated from one hard surface to another. Add up the movements of the parts to get the whole. The function of the whole machine is calculable, and can be anything—a cyclotron, a snow-cone machine. It needn't be known to understand particular parts or trains of parts.

But in an organism, energy flows throughout the whole, and the functioning of the parts cannot be explained in isolation from the functioning of the whole. The whole containing everything is never fully calculable Nature herself—the Whole.[1]

Emerson retrieves a primal or mythic image: circular power returning into itself. The Whole is an organism composed of organs. The parts or organs exist for the sake of the Whole, feed into it, and the vitality of the Whole feeds back into the parts. Since we are organic parts of the Whole, deceiving others, or damaging them in any way, means deceiving and damaging ourselves, obviously or insidiously.

To support this sweeping reversal of mechanistic science, Emerson draws from sources far afield, borrowing a term from Hinduism, Oversoul, to comprehend circular power returning into itself. It is the all-inclusive energy flow of Nature into her member organs and from them back into the Whole. Human cultures are transformations of Nature within Nature effected by human members of the Whole.

Emerson became a popular writer, for he tapped a deep apprehension and longing in many readers. But he scandalized adherents of orthodoxies, including the ever-proliferating ideologists of scientism, and many Christians and Jews, even Unitarians. He reopened a consciousness of Nature that is primitive, pagan, shamanic.

Circular power returning into itself is the power of regeneration. It is "the inexplicable continuity of this web of God" because it is the self-regenerating Whole, dependent upon nothing outside itself, ultimate reality and value, unconditioned, divine. Emerson resonates with vegetables, plants, and trees: "I have no hostility to nature, but a child's love of it. I expand and live in the warm day like corn and melons."[2] And again,

> The greatest delight which the fields and woods minister is the suggestion of an occult relation between man and the vegetable. I am not alone and unacknowledged. They nod to me, and I to them. The waving of the boughs in the storm is new to me and old. It takes me by surprise, and yet is not unknown.[3]

Emerson exhibits a startling ecstatic reach that reveals the intimacy of otherness. He clues us how this happens: "The waving of the boughs in the storm is new to me and old. It takes me by surprise, and yet is not unknown." The present sight is invested with archaic, communal, immemorial experience of it, vague but powerful. Communion of things in space occurs because of communion of moments over time. The past is a living presence.

Immediately preceding the quote in question he writes, "In the tranquil landscape, and especially in the distant line of the horizon man beholds somewhat as beautiful as his own nature." To experience the horizon is to sense that everything else lies beyond it, and that this includes all that was and might be; it is not just everything at this instant.[4] It is a reach in time as well as in space, and, beautiful as our own nature, it draws us into itself. Finding ourselves belonging there, we find ourselves.

In a way that mechanistic physics, caught up in the idea of point-instant reality, could never understand, the past is present. Emerson's futurism, archaism, and startling involvement in the here and now are all of a piece. He offers us a renewed ecstatic reach of self and freedom, and a new approach to how addictions cripple: they split us off from the Whole, and also within ourselves.

Emerson strives to retrieve an attunement that slips from North Atlantic awareness. Peter Nabakov writes,

> As one Choctaw community was about to move from its ancestral Mississippi forest and start the westward trek to Indian Territory, the women make a formal procession through the trees surrounding their abandoned cabins, stroking leaves of the oak and elm trees in silent farewell.[5]

Buffalo Bird Woman, Hidatsa Tribe:

> Often in summer I rise at daybreak and steal out to the cornfields; and as I hoe the corn I sing to it, as we did when I was young. No one cares for our corn songs now.[6]

It is not only dispossessed native Americans who feel this loss; but they can articulate it. They and a few prophetic thinkers of European ancestry such as Thoreau, Emerson, William James, Dewey, in their inspired moments.

The Ultimate Kinship System

We speak to the other and the other—Nature—responds, writes Emerson. "What is it that nature would say?" "The running river and the rustling corn speak of character."[7] They speak of this because they reveal our responsive and responsible openness to them, our courage to let them be themselves as they flow through us. We come home to self only when we come home to other-than-self. Emerson asserts that the woods will be experienced as haunted, as if some activity had been suspended just before we entered, and we were being watched.[8] Immediately involved, self is not subject over against object; one life, one regard, permeates everything.

Everything issuing from any organism-organ—including natural wastes—is nourishment for the larger organism-organ, which feeds into a larger, which feeds back its vitality into its members in the fullness of time. The largest Whole, Nature, feeds back into itself, an ever regenerating, self-sufficient totality. All the dislocations and separations of mechanism—self from other, mind from matter, human from animal, possibility from actuality, present from past and future—are superseded by the most intimate, organic belonging of part within Whole.

Not even the separate moments of linear time are all that needs to be said about time. We must speak of the eternal present. For the vitality of archaic sources is conserved throughout the Whole, just as an individual organism conserves its original genetic structure and its funded muscular and glandular habits. Nature, the organism of organisms, perpetually reclaims its sources and cycles anew into the future, using the wastes of its dying members to propel new life in the Whole.

Emerson: Promise and Problem

Emerson's early thought is arresting and enduringly valuable, paving the way for those ecologists of our day trying to rediscover our place in Nature. For James Lovelock, for example, with his Gaia hypothesis: the planet Earth feeds back into itself and regulates itself very like an enwombing organism composed of members. Despite the fact that the sun now shines twenty degrees hotter on our surface, Earth seems to have maintained a fairly constant temperature: as if it participated in circular power returning into itself, a self-correcting circuit.[9] Although the mechanistic conception still motivates our culture's alienated, addictive, piratical attitudes, Emerson's is a powerful critique.

His vision, however, is overly optimistic and incomplete. As his thought develops, Emerson encounters a side of otherness that resists intimacy—the contingency or sheer repulsiveness of some of the *other*. Particularly the frightened constrictedness of our own bodies resists the fluent circulation of power through us and back into itself. Perhaps we blindly fear our primal needs will not be met. Emerson doesn't know quite what to make of this. Thought transcends ecstatically through horizons, but this is possible only because a body at a place and time does the thinking. But thought finds it greatly difficult to think its own body. He alludes to the repulsive bodily otherness of human beings:

> The transmigration of souls is no fable. I would it were; but men and women are only half human. Every animal of the barn-yard, the field and the forest, of the earth and of the waters that are under the earth, has contrived to get a footing and to leave the print of its features and form in some one or other of these upright, heaven facing creatures. Ah! brother, stop the ebb of thy soul,—ebbing downward into the forms into whose habits thou has now for many years slid.[10]

This might not seem evasive, but I think it is. Emerson is supposing that what is human about us is the half that is not animal, rather than facing the possibility that what is human is humanly animal. He is tacitly using a version of seventeenth-century psycho/physical dualism—the idea of a nonphysical self, or soul, connected somehow to the body.

In another place Emerson speaks of "the divine animal who carries us through the world."[11] The animal is "divine" in this instance, but again the separation of self from body appears: it is upon this *animal* that *we* ride.

Even his "*Over*soul" troubles, suggesting detachment of the animating principle from the Whole, not the immanence of the animating principle in matter. The inconsistency is more than verbal, for it is part of his squeamishness in the face of the human body. This is a bad deficit, limiting the power of what he can tell us about addictions and how to relieve them. For how can we be swept up and integrated in regenerative Nature's circular power returning into itself if we don't accept unblinkingly our own ecstatic bodies and their often concealed or half-concealed ecstatic needs—accept our bodies as ourselves? As it works through us, circular power will be short-circuited addictively.

With regard to race in particular, and the obtrusive bodily reality of it, Emerson cannot clearly see beyond the racism typical of nineteenth-century America.[12] It is *they* who are animalistic. He recoils from close inspection of self as needy body, self as this organic, wilderness spontaneity so close at hand.

A similar bias emerges in Emerson's attitude toward women, and, of course, their distinctive bodily reality. This he sees in a slanted way. And since, according to his own view, body and its world exist in intimate reciprocity, blindness to the body means blindness to the whole world as known by that bodily knower.

> Few women are sane. They emit a coloured atmosphere, one would say, floods upon floods of coloured light, in which they walk evermore, and see all objects through this warm tinted mist which envelopes them. Men are not, to the same degree, temperamented; for there are multitudes of men who live to objects quite out of them. As to politics, to trade, to letters, or an art, unhindered by any influence of constitution.[13]

Not even Emerson's close relationship with the remarkable intellect and writer Margaret Fuller could free his views. Did he think that even she counted as one of the few "sane" women? Fuller, like many gifted women, was badly impeded by self-doubt, and as her friend, Princess Christina Belgiojoso, said about herself (and Fuller might have said about her own self), "She had spent her youth 'like a rat in a library when she was allowed to do as she chose, and like a doll in the parlor when she was not'."[14] The disgust with a bodily reality that enters a forbidden "spiritual and intellectual" male domain is clear: like a rat in a library. Women—so material, so organic, oozing—may transgress boundaries; they are profoundly disturbing. Behind their conventional facades they are uncanny, taboo.

The blindness and intensity of Emerson's emotions suggest that projection is at work. Fear, anxiety, and revulsion toward our own body is too much to bear, and the problem is thought to inhere solely in those body-ensnared others—women, other races, wild animals. Then to protect ourselves (particularly white male body-selves) from the nightmare creatures we have created, we wall ourselves up in addictive "self-sufficiency." Emerson's famous idea of self-reliance must be carefully delineated to exclude this perversion of it.

It was a cold winter day in the fields. The ground was not yet iron-hard, but crystals of ice had formed between clods of earth and crunched slightly under foot. Jets of steam shot from the dog's nostrils and shone silver in the sunlight. He walked gingerly on the rutted road. Seeing the ghostly herd of deer on such a cold day was unlikely, but I anticipated them nevertheless. The dog and I were scanning the horizon. I hoped at least to get to the clearing in the woods and drink the hot chocolate that I carried in a small thermos. We walked past the house site but did not stop: I merely glanced at some fragments of blackened brick which at one time formed the chimney of a home.

Where the road turned to skirt another field stood some hillocks of dirt, pushed there, I supposed, by developers who had cut a broader passage into the field to prepare it for attack. These had stood for years, awaiting further action. Up one of these mounds we walked and surveyed the fields and woods around us. The air was transparent, still, odorless, as if the sun glistening on the frosty landscape were jealous of any other sense distracting us from its spectacle. The horizon was etched cleanly in the distance. Close by, on our right, stood browned and broken bracken, where several months ago, in the fall, a large doe had bolted up only forty feet in front of us and bounded across the fields. Now all was immobilized in the sunlight.

Then my vision was caught by something moving in the distance, coming out of the woods over a hundred yards away. It was little more than a moving dot, but loping in a peculiar way. Fixing my eyes on it, I tried to discern what it could be. It kept running straight at us, and though hypotheses flicked before my mind as rapidly as cards rippled in a deck, none matched and characterized it.

It kept running toward us, and a slight uneasiness joined a mounting fascination. It was not a dog, a cat, a ground hog. Then I made out in a flash what it was: an animal I had never seen before in these fields, a red fox, its bushy tail flowing out behind it as it ran toward us still. Now it was only about fifty yards away. Suddenly I was seized by panic. What if it would not stop?

I had no idea what the dog would do. Buppie looked uneasy too. I began to shake. My own fear startled and frightened me. The fox kept running toward us.

Then the animal paused for an instant—I do not know if it had seen or smelled us—and changed its course ninety degrees, and finally circled back toward the woods.

I was fascinated and strangely elated; but my fear stayed in the background as I walked with my dog back home.

Varieties of Fear

How could a grown man, not unacquainted with wilderness, be frightened by a small wild animal? But it was a fact. Unattuned to wilderness in those moments, I was unprepared for its approach. Fear was unclarified and confused. I failed to respond competently and ecstatically.

Might there be a fear that is awe, a kind of holy fear that might stop me dead in my tracks for a moment, but allow me to continue? A fear of the encompassing everything-else, of the World-whole of which I am a small part, a fear that is reverence, that makes me belong? A holy fear, simultaneously sacrifice and sustenance, a losing of nervous ego-self that leaves room for something much greater?

Holy fear lived through is world-excitement, part of which is the lure of not knowing clearly what holy means, how the vast teeming world can flow into us and support us, or can crush. Who knows? As a leaping fire sucks in air, holy excitement feeds on faith.

As Black Elk told John Neihardt,

> When I was still young, I could feel the power all through me, and it seemed that with the whole outer world to help me I could do anything.[15]

Being with the dog was often sacramental. The more I allowed him to take the lead—the more the world around us flowed in and took over—the more coherent and ecstatic I became. But I did not know, could not remember, how to turn the encounter with the wild animal into a sacrament, or that abandoning that is a finding. So we stumbled home, oddly elated but stymied.

Sacramental Power

Ecstatic reach points toward the unknown, untouched, spontaneous, and unspoiled, the wilderness to be explored beyond the boundaries of human habitation and technology. And it affirms the presence of the unknown within our bodily selves, the whole array of alerted organs that are typically never seen. It is the unknown and untouched wilderness within, the beyond within, that moves so incredibly intimately with the beyond without, and the without feeds back into it. The wildernesses interdigitate and interfeed and move us some way emotionally.[16] The dark, inner body-self resonates with the world, in various rhythms at various times. The bright days have their rhythms, as do the nights, and the winter its rhythms, and the spring its. Letting regenerative Nature have its way is sacramental.

Many of us today are both attracted and repelled by what wilderness might bring forth. Unclarified fear predominates. What if the separation between self and other should blur, and we break into the unclassifiable, the uncanny? How could we learn to identify and accommodate holy fear, world-excitement? If we are not at home in wilderness source, we can acknowledge only a blind fear of abandonment and of bodily harm. Yet much more than this is dimly felt to be possible, so confusion ensues, sometimes panic.

Today horizons are crossed and set resonating by limited resources—for many by casual sex or unritualized drugs or by violence—and we must strain to articulate the resonance. The modern conception of mind as self-inventorying detached consciousness places us at a miserable distance from periodicities and spontaneities of wilderness Nature, her stresses and relaxations, to which we belong.

It was not always so. As Hans-Peter Duerr tells us in *Dreamtime: Concerning the Boundary between Wilderness and Civilization*, before 1450 and the three centuries of witch-hunting mania that followed, each village had "our witches," the *hagazussa*, those who "sat on the fence" between wilderness and civilization and eased and regulated passages across it and back again. This name has been shortened to *hag*, and it means crone or wise woman, witch.[17]

It was only when Christians became especially defensive in the very late Middle Ages, during the rebirth of pagan studies and the upsurge of modern secular science, that the hags were considered dangerous and were hunted down and killed by the hundreds of thousands or millions over three centuries. And they were dangerous, at least for societies that had polarized the world so rigidly into present versus past, spirit versus flesh, heaven versus hell, men versus women—with women so easily falling from their angelic "good woman" perch into the "bad woman's grossest fleshly lusts and animality." (Goddesses of every epoch have been associated with particular animals as well as plants and trees.[18])

But there is always something about the boundary or horizon that is dangerous but essential for our vitality. So the incalculable interfusing must be lived through, endured or enjoyed. Attention and concern, love and hate, place us beyond ourselves ecstatically, beyond our full control. To try to suppress or conceal the danger is to become manic and hysterical, or dull and empty. When primal ecstatic needs to cross horizons are unrecognized or repressed, counterfeit urges emerge and often become addictive.

Emerson frequently represses dangerous resonance to wilderness. Those hags sitting on the fence between wilderness and civilization frightened him, I think; their bodily reality and the wild resonances that go with this disturbed him. We identify with things ecstatically, or, threatened, exclude them to our impoverishment.

Openness and Vulnerability
Brilliantly Emerson revealed our ecstatic openness to the world. But he failed to see a corollary of this: we are vulnerable bodies whose identities are easily threatened by the world around us, even by our own organic processes, particularly today with the loss of hunter-gatherer skills and rituals that integrated us in the Nature that formed us. It is comforting to think that our true selves are pure minds or souls somehow ensconced within, and buffered by, our bodies. But I think the self *is* the vulnerable body.

The greatest primal need and satisfaction is to expand ecstatically in imagination and deed into will-of-the-place, domain after domain, and to be confirmed as a whole individual within the Whole. But being vulnerable, we may contract into a shell and try to be satisfied with substitute gratifications. And these, because ephemeral, must be addictively repeated. The effort to avoid the anguish of withdrawal is itself anguishing. The urge to expansion, freedom, sound individuation—profound enjoyment—collapses into servitude.

To fail to grasp the body and its primal urges as body-self's is to identify with only a fraction of oneself—the pure will or immaterial soul or intellect. Believing the self and its mental powers are inside the fortress of the body engenders smugness and a semblance of control and security. But behind this fortress mentality is unclarified fear—along with body-self's primal needs disowned. Substitute gratifications throw us into impotence and confusion. Dewey:

> Oppositions of mind and body, soul and matter, spirit and flesh all have their origin fundamentally in fear of what life may bring forth. They are marks of contraction and withdrawal.[19]

Emerson's lucidity and brilliance do not go all the way down. A kind of drugged drowsiness appears at a certain level, a repression of body's disturbing needs and possibilities. He does not help us integrate ecstatic bodily urges and make them fully our own.

Emerson's preoccupation with the Oversoul and his elevation to divinity of circular power returning into itself distracts him from the always locally situated body, its vulnerability and dependence on others—particularly on that woman within whose body each first experiences the world, one's mother. Because of this, his account of Nature and its educative force, brilliant in many ways, is partial and abstract at a crucial point.

When dependency and vulnerability are not viscerally acknowledged, we cannot effectively distinguish fear of bodily harm, or various forms of childish and neurotic fear, from holy fear of the never completely comprehensible or controllable World-whole that has given birth to all things, including our own organisms. We disengage from the most ancient myths that humanity has devised to consolidate our ecstatic but at the same time always localized condition: from Mother Nature, or, in Paleolithic times, Earth Goddess.

Numerous images of her carved in stone or bone, go back at least thirty thousand years to the roots of culture in human organisms' transformations of Nature. We live ecstatically, stretching beyond spatiotemporal locality, but always within a womblike matrix of matter and energy that flows into us and supports us beyond our ability to comprehend. Or which disturbs and frightens when we attempt to deny this never fully predictable reality.

The pinching off of cycling spontaneity, and the absence of any acknowledgeable sense that anything is missing, do not merely result in a nebulous restlessness "in the mind." Body-self, frustrated, crashes about—or quietly withers away. Again, Thoreau early in *Walden*: "The mass of men lead lives of quiet desperation."

The only Australian cities of any size lie on the coasts of the continent and look outwards toward the regions of the world from which the European settlers came. Though their ancestors arrived as early as two hundred years ago, many or most European-Australians have never visited the interior of their land. One spoke of the central Outback region as behind the beyond, another referred to it as the dead heart of the country. Tales of travelers' disasters—many of them true, no doubt—were woven about us frequently. One spoke of poisonous tiger snakes that stalked human beings. When I told the Aborigine Burnum Burnum how

*surprised I was that so few Whites had visited the interior of their country, he
said, "It is because they do not know the interior of themselves."*

The innards of the body and the innards of the land one system? Bizarre, as well as strangely disturbing. The Aborigine's words haunt. Facile divorcements of mind from body, self from other, organic from
inorganic, present from past, and obsessive preoccupation with vision
at the expense of the other senses, particularly the kinesthetic and tactile, alienate and fracture. We may look about nervously and turn our
faces up—like children frightened by things that loom above them.
What lies closest to hand and under our feet is overlooked: body embedded in the pulse of matter that surrounds, pervades, supports or
threatens us.

Notes
1. Again, words referring to the unconditioned and unconditionally valuable
   have mythic force and are capitalized.
2. "Nature," 70.
3. 39.
4. Compare David Bohm on domains not spatio-temporal: interviews with
   Weber, in Renée Weber, *Dialogues with Scientists and Sages*, 39–41.
5. *Native American Testimony*, 151.
6. 182.
7. "Nature" and "History," 43, 152. For a contemporary Emersonian reading, see Alphonso Lingis, *The Community of Those Who Have Nothing in Common*, particularly "The Murmur of the World."
8. "History," 158–159.
9. Lovelock, *Gaia*.
10. "History," 167.
11. "The Poet," 274.
12. As Cornel West has brought out clearly in *The American Evasion of Philos-
    ophy*.
13. In Cornel West, *The American Evasion of Philosophy*, 247.
14. Paula Blanchard, *Margaret Fuller*, 283–284.
15. Neihardt, *Black Elk Speaks*, 180.
16. For this revealing spelling, e-motion, see Glen Mazis, *Fragile Ontology*.
17. Duerr, *Dreamtime*, 65.
18. Buffie Johnson, *The Lady of the Beasts*.
19. *Art As Experience*, 22.

Chapter Four

The Intimate Otherness

of Body-Self's World:

Addiction As Frightened Response

"Body am I, and soul"—thus speaks the child. . . . But the awakened and knowing say: body am I entirely, and nothing else; and soul is only a word for something about the body. . . . "I" you say and are proud of the word. But greater is . . . your body and its great reason: that does not say "I," but does "I."
—FRIEDRICH NIETZSCHE, *THUS SPAKE ZARATHUSTRA*

Each of us is encased in an armour whose task is to ward off signs. Signs happen to us without respite, living means being addressed, we would need only to present ourselves and to perceive. But the risk is too dangerous for us, the soundless thunderings seem to threaten us with annihilation, and from generation to generation we perfect the defence apparatus.
—MARTIN BUBER, *BETWEEN MAN AND MAN*

"Just say No to addiction." This advice is well-intentioned. Moreover, a sense of our ability to choose and to make a difference in our lives satisfies a primal need: to feel ourselves to be significant.

In many instances, however, the injunction is futile. It assumes there is some self (pure self as mind?) that could step outside of its body and order it about. But if selves are body-selves, this is delusion.

The delusion divides body-self within itself, and promotes addiction of one kind or other. For when body-self's primal needs and urges are regarded as "physiological events to be conquered," they cannot be owned and accepted as self's. They take on a life of their own in some rebellious, counterfeit form. They are mockingly both "me and not me"—"me" insofar as they can't be understood in the detached way we grasp a physiological event in our bodies *as* such, but are productions of body-*self*. Yet also "not me" insofar as they defy my will and agency.

This experience is disorienting—we move in a kind of trance, as if possessed. And, indeed, addiction as a form of possession will be taken up in chapter 7.

It was six o'clock on a winter's night and I was to face 240 undergraduates in ten minutes. I was listless and completely empty. The urge to walk to the "grease trucks" to buy chocolate moved within me, a tidal force. Seriously hypoglycemic all my life, I knew perfectly well that I would be high for twenty minutes, then plummet, be empty, jittery, with a burning stomach, for the remaining hour with the students. I got out of my car and walked toward the trucks. I asked myself, "Are you in charge of your life or aren't you?" I kept walking toward the trucks. Now only one hundred feet away, my face was prickly and hot. I kept walking. I had done such things for decades. Then, as if taking myself by the neck and shaking myself, I broke through, and forced my steps off the sidewalk and across the street toward the classroom.

The next weekend at a party a woman asked me what I did. I said I taught in a philosophy department. She asked, "What are you working on now?" A moment's panic: I had to tell her exactly and precisely that it was addiction I was working on, and how an addictive craving is ambiguous, is "both me and not me," I heard myself blurt, "it is, it is . . . a trance that has to be broken through. It is a TRANCE!" The relief was indescribable. She seemed to understand me.

About another week later, I recounted these incidents to Glen Mazis. "Is it in your book? . . . It should be." Obvious, once he said it, but I had not thought to do it! Or could not, not so publicly and finally? Immediately I wrote what you've just read. Ezekiel 1:1–18:

I saw visions from God . . . A strong wind came out of the north . . . and out of the midst of it . . . glowed amber metal, out of the midst of the fire . . . As to the appearance of the wheels . . . a wheel within a wheel . . . As for their rims, they were so high they were dreadful . . . and they were full of eyes round about.

Addicted life is a wheel that cannot imagine its limit—dares not imagine the wheels containing it, their rims of eyes reading and opening us to all, but which we cannot read. In what may I still be trapped?

Addictive Cravings and the Blur of Guilt

If addictive cravings could be objectified as merely physiological events, they would not excite guilt. They might prompt anger, frustration, disappointment, but not the guilt they typically do excite. Guilt implies some responsibility for the condition, and this implies that *self* is at work on some level, the *bodily* level, not merely some mechanical or impersonal force.

But what of this objection: "I may feel guilt over what's mine without its being me. I may feel guilt over my dog, if I should have kept him in his pen, and he terrorizes a neighbor's dog. Indeed, I may feel guilt over my own body without its being me."

This will not do. If my child dies, say, something profoundly mine has been destroyed. There is great loss for me. Space shrinks, it becomes a mere cavity. Atmosphere moves against my body, but it is hardly wind—it is dead. I might take my life.

But I must be alive to do it. What is uniquely and indispensably mine must remain: my own body—that without which I would not be who I am. What my body does, I do. If it shoots, I shoot. If it faces north, I do. If it walks, so do I. Nothing else is so uniquely mine that it is also me. It is profoundly mine, but more—it is me.

What must be explained is guilt over addictive cravings, and how guilt is projected onto others and other things so self-deceivingly that I am usually powerless to break through it. This profoundest self-deception and evasion presupposes the profoundest guilt, that is, guilt over *who I-myself am* and what I have done, not merely over what something that is mine is, or what it has done. Generating cravings must be violating myself, and this is why it is so painfully difficult to admit what I have done. Though I may deceive myself about my dog, this is typically not as self-occluding and intractable as that concerning my cravings. Or though I may deceive myself about a bodily condition that just happens to me—a disfiguring scar, say, or a tropical disease—these self-deceptions are not as self-occluding and intractable, because it is not I violating myself. It is a bit easier to get some distance in these cases.

To explain the profoundest guilt and its projection onto others we must suppose abysmal confusion amounting to trance: confusion within myself. Disoriented, frightened, ashamed, addicts freeze in repetitive cravings. It must be that the appetitive body as the self has the potential to derange itself: to appear to itself at inappropriate times as thing-like or thing-y in its cravings, as "not me" *as well as* "me." In the face of ambiguous and disorienting cravings we allow ourselves to be entranced, much like, once in bed, we allow ourselves to fall asleep.[1] Soren Kierkegaard puts it brilliantly and cryptically: freedom swoons and arises guilty.[2]

Deceiving ourselves we trade on the ambiguous experience, and nudge the "me and not me" experience to the "not me" side: "Others or other things cause my problems." In prereflective rush, the nudge is recognized by body-self on the margins of awareness as "not to be recognized further," for if it were, I would have to acknowledge my guilt. Our spontaneity does not simply exclude freedom and responsibility, then, however confounded, miserable, and pathetic they are. How can we admit to our freedom abusing itself, to ourselves as agents abusing ourselves—how admit to failing ourselves at the heart of ourselves?

Not all addicts self-deceivingly project guilt, but most do. Unprepared by our culture for the intimate otherness of things, and the even more intimate otherness of our own needy bodies, it is easy to blame everything but ourselves. It is "what one does."

Western logic contributes.[3] The principle of noncontradiction—something cannot both be what it is and not what it is—impedes awareness of the real ambiguity of cravings, and tends to limit our dealing with them to mendaciously repressive and projective or stupidly permissive ways. This logic is a tool essential in many contexts, but habitually applied to oneself turns on the wielder, fans the flames of self-deception, and burdens those upon whom guilt is projected.

Only in a whole world in which we loose ourselves competently and ecstatically can we be coherent and powerful. But to "let go" after losing our primal hunter-gatherer skills seems either counterintuitive or terrifying. How to react sanely to Emerson's prompt: "The one thing we seek with insatiable desire is to forget ourselves . . . to do something without knowing how or why"? This would not be murky trance or evasive projection, but the keenest wakefulness.

Hysterical fear and confusion is withdrawal into "worlds," constricted rounds of substitute cravings and satisfactions, the projection of guilt, the wheels of addiction. How do we wake up and slip free?

The Bodily Self

Given the habitual tendency to think that self is divided into mind and body, with mind as the heart of self, buffered from the world, addiction can never be understood. That is, if self is conceived to be *my*self—an essentially self-reflexive and self-appropriating being—and mind is consciousness conceived to be essentially awareness of itself *as* itself— then it follows that mind *is* the self, or the heart of it. But to think that mind is a domain separate from body is false, and it cannot explain addicts' guilt.

Dualistically conceived, guilt is inexplicable. Addiction is thought to be a brute physical force that overwhelms the mind—but why should *I* feel responsible for that? Or it is thought to be the work of the devil— without grasping what demonic possession is.

Now the challenge of questioning systematically the reigning dualis- tic assumption, of taking Nietzsche and Buber seriously, of grasping what *body-self* is. This means knowing what primal needs and potentials are, and what mature body-selves are. These persons identify fully with primal needs and what meets them.

But "body-self" involves more than merely carpeting together two words. Indeed, "body-self" by itself is a snippet, an abstraction. It must be "body-self-world." The degree of maturity of body-self is the amount of the compatible world it makes its own ecstatically and responsibly (recall swimming confidently in the ocean: the waves are one's own in the very abandonment to them).

I do not think the self can be understood without understanding the body as we live it personally and immediately moment by moment. So lived, bodies are not objects, but ourselves caught up ecstatically in the world, for good or ill. If we understand the self as this living and experi- encing body, we won't think that trying to "step outside our bodies" and saying No to our impulses and hungers will be sufficient to curb them in all circumstances. Because our very being as selves includes the body's powerful addictive cycles, its weirdly imperious cravings and appetites—an uneasiness in the hands, a gnawing emptiness in the stomach, an urging in the groin. Addictively repeated acts are body- self's frightened, feckless attempts to preserve its continuity and iden- tity, without fully acknowledging its bodily and needy nature, and without happily accepting its multidimensional ecstatic involvement in the regenerative world.

Perhaps no ecstasy generates more seductive allure than standing out perversely from one's body in thought and denigrating and trying to dominate it. If disgust with the body is too much to bear, it is projected ecstatically onto others and their disgusting bodies. William James described the body-disowned as "the excrementitious stuff, the sorry remainder of the dualized self."[4]

Being a body-self! We must rethink a whole cluster of questions—of development, responsibility, identity.

Living the Body-Self

We can never completely objectify our own bodies even if we try. If we look in mirrors we can see only one section of our surface at a time. Moreover, we have prepped ourselves, "put on a face" (and a body) for the occasion. Candid camera shots are better, but still the objectification of the body is not total. If we forgo mirrors and usual modes of objectification and try to catch ourselves out of the corner of the eye in the very act of acting, thinking, feeling, we detect a portion of the body coming into the focus of this reflecting consciousness, and it appears as having been already in the world, and alive, before it appeared to reflection.

In other words, in experien*cing* of various sorts (perceiving, hoping, thinking, praying) my body can be experien*ced* or objectified by me to various extents at various times. But at every moment the body is caught up in experiencing and never able at that moment to reflect and objectify all its experiencing. There is always some of it unobjectified; to try to objectify a moment's experiencing requires a *later* moment and *its* unobjectifiable experiencing. The present moment in its very presentness can never be completely reflected, known, controlled. A primal spontaneity inheres in the experiencing, thinking, feeling body— though it may be scotched when it begins to be acted out. This spontaneity is at the root of our primal need to explore, grow, and be.

The "engine" of my being, that which I can never get behind and objectify totally, that which drives my life—my "inner" "subjective" self—might be my living body just as I live its totality? The visceral needs and feelings that drive the healthy unaddicted self . . . are they so much *this self* that they cannot be neatly individuated, objectified, set out in front of the self to be studied and managed?

But denials spread like wildfire. Imagine this one: "We cannot conceive of bodies experiencing—thinking, seeing, fearing, hoping. *Selves* must do this because they have *minds*. A self thinks of itself remaining itself as it passes through multifarious experiencings. It can do this because it has a mind: things of endless sorts located at endless times and places can be thought of all at once because they can be thought together in one thinking, in the one abiding mind that makes me me. To suppose the body does all this is absurd."

Gathering together things experienced from hither and yon, of no matter what sort, and experiencing them all as my own self's experiencing of them now, *is* wonderful. But to say it is the mind's work and not the body's divorces us from body and its needs and world, from body-self, and leads directly to uprooted freedom, addiction, life alienated from itself and the rest of Nature. To speak of thinking and feeling happening in the mind distracts attention from body-self's thinking and feeling, from its and my urges, hungers, spontaneities. It invites the "me and not me" equivocal self and the probability of moving in addictive trance.

The expression "the mind" is understandable but regrettable. Like "the skull" and "the brain," it leads us to expect some kind of object or thing. Finding the mind-thing to be unobservable, unlike the skull-thing or brain-thing, typically prompts belief that it must be a nonphysical or spiritual thing.

It would be much better to say "minding." Though it sounds awkward, why couldn't the body do the minding? Then it would not be excluded from self's feeling and thinking by the mere language we happen to use. We could see how body essentially involves self and mind, yet not expect to be aware of all, or nearly all, of this involvement. For we have no direct awareness of our nervous systems and the great knot within it, the brain.

> Thoughts and dreams light as
> Air come as if from nowhere.

Neuroscience makes considerable strides, but its findings do not allow direct translation into the immediate, crude, basic feelings and terms we use to make sense of our lives. Evidence does indicate, however, that the brain's functioning is essential for minding, experiencing, and the sense of self. For when the brain is severely impaired (as in Alzheimer's disease, say), the victim's minding and sense of self are severely impaired as well. Of course, there might be a nonphysical mind-or-soul-entity, undetectable by science or ordinary experience that is merely connected somehow to the body. But it is hard to imagine what it could be.

We can have direct awareness of a portion of our bodies. Sometimes this awareness is focal, as when we have an upset stomach or stubbed toe, for example, and play physician to ourselves. Most often, though, the body appears to us only in the fading-out margins and not in the focus of consciousness, and so is not neatly demarcated or objectified. It has the capacity to be an experiencing-experienced being, while things other than it would appear to be only experienced (or experienceable).

When we free ourselves from the prejudice that body with its urges is just an object, a thing perceived "out there," we begin to see how what we can describe of it as each of us lives it is essential to grasping minding and sense of self—essential though not perhaps sufficient.

Bodily Roots of Self and Mind

Let us suppose that as the body moves about in the world—following its needs and impulses, overcoming inertia—the organism experiences a contrast between the constancy of its experience of itself (mainly marginal) and the inconstancy in changing circumstances experienced. Moreover, the movements of the body and those of the rest of the world typically change independently, or one remains constant and the other changes. Since the body's ability to move itself about seems to be a fundamental fact that the person involved needn't explain, why not suppose that this is the root—though not the trunk and branches—of the sense of individual self as agent? Perhaps this is the germ of the feeling, "This being who does all this and remains itself and has these experiences is just I-myself." A limited claim, but awakening possibilities for living as a whole self, for reweaving appetite, will, feeling, need, understanding, and for grasping how they can fragment addictively. If minding must involve bodily activity, to try to go out of one's body by manipulating it addictively is to try to go out of one's mind.

However, it still sounds absurd to say "My body regrets doing X," "My body hopes for Y," "My body thinks of a quadratic equation." But it sounds so only because ordinary language assumes the body can only be an object, one studied by physiology, say. But this just assumes that the body cannot be both subject and object (object for itself to some extent sometimes). This just assumes that the body cannot be both "owning" experiencer (with its experiencing)—"I"—*and* owned, experienced—"me-myself" (in its least form, abstracted from the world). The structure of common language merely assumes that the body cannot be body-self, I-myself. We shouldn't be satisfied with that! If the structure of common language had evolved to aid people to find their way out of addiction we wouldn't have so many addicted people.

Let us further unfold the rudimentary structure of both minding and body-self: different things from different times and places are thought about or experienced within the unity of one thinking or experiencing that contrasts itself as experiencing-experienced being to what is merely experienced. They are *other*. This self-aware body-self would transcend disparities of time and place and sequences in what is experienced by it, and in doing so would experience itself remaining itself through all its experiencing—the source of all its actions and attitudes and other than all its others. It would be the self—body-self.

In *Beast and Man* Mary Midgley suggests sagely that the greatest need is "to maintain our continuous central life," the sense of I-myself through time.[5] I believe this happens when the ecstatic body-self forms itself maturely: alternately, in the flickering of an eye, fuses with its surroundings, then differentiates itself from them. The fullness of the life maintained is a function of the scope and depth of both ecstatic differentiation and fusing. This is a corollary of the general theory of value: the greatest differentiation and individuation within the greatest coherence and unity.

What do we mean exactly by fusing or merging? Since in spontaneous, prereflective life the body is not objectified in the focus of awareness, we can be caught up ecstatically in other things, sometimes carried away. But, you may say, what's seen is always seen from the point of view of this body, whether it is in the focus of awareness or not. What is heard is heard from "the point of view" of this hearing body, whether the body is focal or not, etc. All the presences of the world present themselves here, at this movable site.

This is a thoughtful point, but it must be heavily amended. Such a reading of the body is influenced by the Cartesian assumption of a separate individual "mind in here" in which images of something that's perhaps "out there" assemble. But when we look freshly at our actual living we find that this mobile center—I-myself—often gets radically peripheralized. We can be so caught up in things, so magnetized by them, that this allegedly ubiquitous "point of view—me!" is not only not focal in experience, but is not even marginal in a behaviorally significant way.

And if we do not catch ourselves, that is, make our individual bodies more or less focal, we will get carried away in constricted and addictive groups, in contagious feelings of various sorts—in celebratory revels, fads, indulgences, perhaps something like St. Vitus dancings, or lynch mobs or pogroms.

When we look freshly at what actually happens in our mergings, we see that our experiencing is not completely isolated from others' experiencing, whether human others or not. Though we may not know exactly what others are thinking, just as they are thinking it, we do get directly the tenor of it, and its vibratory reality moves through us.

But this means that my initial characterization of individual identity as "I am an experiencing-experienced being while others are only experienced (or experienceable)" was too simple. The fact is, how I am placed and defined and caught up as a member of groups, how others' experiencing feeds into mine, is essential to my identity. Human body-selves are vulnerable, permeable, strange.

Corporate Bodies

Yes, a private aspect of our body-selves is undeniable and unspeakably valuable: particular bodies' immediate kinesthetic and kinetic experiencing of their own dark insides. But as we shall see, even this private aspect gets expressed to others to certain degrees in certain ways in certain situations. To be sound and mature, the whole body-self must feed ecstatically into the vast surrounding world and get a confirming response. But, we will note, the group in which it is immersed may itself be addictive.

A crucial sector of my circumstances are other humans who recognize me to be a human because I have an alive and aware human body. I am caught up in their recognizing, their experiencing, mimetically engulfed in them. That is, I imitate their recognizing involuntarily and immediately, and so completely prereflectively that I cannot acknowledge at the moment that it is *I* recognizing *them*, or *they* recognizing *me*. All this occurs on the dim margins of body-self's consciousness, or out of consciousness altogether.[6]

Others' recognition of me—particularly in childhood—becomes my own recognition of myself. Others form my social self. As appropriated by me, their images of me belong in my body, says James. And most formatively when I do not distinguish their contributions *as* theirs.

The directness of this connection can unsettle. For example, when reproved by an authoritative other, my feeling of shame "is simply my bodily person, in which your conduct immediately and without any reflection at all on my part works those muscular, glandular, and vascular changes which together make up . . . shame."[7] True, as adults we are very often aware of the other as other. But the openness, vulnerability, blurred identity of the infant remains to some extent throughout life, though the mature person can usually muster a somewhat detached and critical response. Shame is a negative relationship, of course, but it exemplifies how the contrast between self and other can blur. Since an other's presence can haunt us in a kind of timeless way, the other's reproving voice can assume a constancy that becomes indistinguishable from our own. We live stunted in unclarified, neurotic fear, or in numbed addiction.

Mimetic engulfment is a corollary of the marginality of our bodies lived on the fringes of consciousness, and their being caught up ecstatically in the world. Other beings are undeliberately imitated through absorption of their presence into the body. Whole groups are absorbed, corporate bodies, or we absorbed in them. Note again the contagiousness of fads, of rapacious consumption, say, that sweeps across the planet. Responsibility is short-circuited in a kind of waking dream.

Something else prompts absorption in corporate bodies, very probably: memories of the womb, now untraceable, in which the self-world relationship was exceedingly blurred and inarticulable. We can speak of muscle memories.

Mimetic engulfment easily conduces to addiction. Take an authority in the primal group whose voice sounds in a kind of timeless way, Do this! Do this! Don't do that! Don't do that! "Old tapes from the past," as some psychologists say. The voice is indistinguisable from one's own. But it's not one's true voice as a mature person now. Again we recognize stupifying ambiguity: in this case an ambiguous voice which is like an addictive urge that is not clearly "me," but is not clearly "not me" either. The disorientation is a kind of panic that inclines to rigid repetitions and blocks the creative flow of development.

Reason and Choice

It's not that powers of detachment and of reflection and choice are never present. On many pressing occasions the mature experiencing body-self can contrast itself to what is experienceable by it, what is other than it. It can pull itself together over time and take initiatives and hold itself responsible if that is appropriate. We feel, and may formulate, ethical principles by which we judge our own instincts and urges (work involving one's own unperceivable brain, no doubt). Our reflexive sense of ourselves as whole, as a self that can be conscious of itself, seems to require an experience of reason confronting at times impulse and desire. And we also need a clear distinction between thought and deed.

Indeed, without addicts' ability to morally judge their urges (at some moments and at some level of awareness) they could feel no guilt, and without this there would be nothing from which to escape through evasion and blaming. They would not be who they are.

But the pervasive tendency to equate the self with that part or level of us that in bright awareness takes our selves to be autonomous, coherent, solid, and articulably responsible—that is oversimple. Grand it sounds, but uprooted and addicted freedom is suggested, willfulness.[8] In fact, many are not mature, and most of the mature have bad moments.

Particularly in North Atlantic culture, people are inclined to grossly overestimate their autonomy. Engulfment in groups that teach that we are not really our bodies especially inclines to addiction. Or ones that teach that our bodies are insignificant relative to the group body and its destiny. Or groups that condition their members to believe that they are superhuman and others are subhuman. All this distorts recognition of primal truths and primal needs: each of us is a body-self open and vulnerable in the world; and, with respect to primal needs, humans everywhere are more similar than they are different.

Body-self is continually challenged as it adjusts and readjusts to the shocks and enticements of the world, as it tries to maintain a tolerable balance of otherness-in-this-body—particularly of group-identity-in-this-body. The gross disparity and tension in North Atlantic culture between allegedly supernal autonomy and actual group engulfment is remarkably acute. Body-self can derange itself.

I had come to the university years ago to chair a small department of philosophy in an unusual college. It conferred bachelors degrees, but its classes were held at night and most of the students were older people, usually in their late twenties or thirties, who worked in various bread-winning occupations during the day. The job offered many advantages: older, serious students, easily manageable administrative chores, reasonable teaching loads, proximity to New York City, decent salary.

A drawback was obvious: about half the courses were taught by coadjutants, part-timers, poorly compensated instructors. As if to compensate for this fact, a number of full professors and chairs had national reputations—albeit only some in the academic mainstream. Others had made it as individualists. For example, colleagues included an economist whose talk was not awash in mathematics, who could converse about Dewey, and a political scientist who lectured broadly in various languages in third-world countries.

It took me awhile to see clearly another drawback to the appointment, one that I should have anticipated. The dean of the graduate school did not bother to look at me when I met him and shook his hand. (The dean of our college told me that he could not get past the secretary in this graduate dean's office.) Sending a copy of one of my books to an eminent professor whose field overlapped with mine brought no response. When I gave a talk, no professors of philosophy from any of the other colleges came.

This situation puzzled me for awhile. My response flickering through a broad spectrum, from indifference—feigned or real—to contempt, anger, amusement. But the answer finally came crystal clear: as professors in other colleges were sitting down to dinner or settling into their second round of drinks before dinner, we professors in my college were beginning or continuing our day's work. That is, we were behaving as servants do. We were stepping and fetching. Reputations outside the university made little difference. Within the corporate organism of this university we were low, not actually untouchables, but precariously close to that status.

As time went on, I felt more angered than amused. In retrospect now, I think I was also frightened. The fear was of a confused, unacknowledgeable sort, as if I were engulfed in a corporate body the outlines of which I could only guess. As if on some dreamlike level I were lunging and plunging, trying to escape, gasping for air.

Addicted and Frightened Groups

Our identities as body-selves may pose an intractable difficulty: to become sound individuals we must be recognized as such by our local group. But this group's identity may be so vulnerable that it individuates itself addictively from other groups at their expense. There can be electrifying ecstatic involvement (recall waves of tens of thousands of upraised arms saluting Hitler), but it is addictive and febrile, not regenerative for long. To show allegiance to our group tends to divide us from the source that formed us, the World-whole, and from our own organisms.

As we will see ever more clearly, the mature body-self believes itself to be a member of a world-community of beings, and though it deeply desires recognition and respect from others in the local environment, will not buy this if it means denying allegiance to the ongoing Whole. Socrates is the classic model here. Mature persons are pulled toward the realization of themselves as individual, social, and world-communal beings.

There is a greatly underutilized resource for identifying with the World-whole: the Other that all cultures share—the intimate and ultimate Other—Nature. Its underutilization is no mystery. Technologically advanced urbanized countries close in upon themselves addictedly. Fearfully at bottom, for they dread their fragile and porous identity be polluted by anything alien to their comprehension and control. They exploit others for their natural resources ("raw materials— mere matter") to be regulated and worked up by North Atlantic computers and machines. "Buy cheap, sell dear."

Precarious Identity and the Body

In the view being developed, identity of self as body-self is complex and perilous. If self is essentially bodily, self will be perpetually open to the world and vulnerable, for body is.

William James had a genius for disclosing the instant of actual encounter with the surround: the often shocking and piercing actuality that words and abstractions, coming a moment too late, typically miss. Only actuality can satisfy actual primal needs. Yet actuality of a different signature can frighten into addictive constriction: for example, if we fear our needs will not be satisfied, they themselves will frighten, and will shrink from the focus of experience into the margins where they cannot be clearly recognized and satisfied. We may self-deceivingly deny their existence.

James notes that what's immediately perceived may be experienced as a naked blob-like "that," an indeterminate something or other, not yet any definite "what." For example, I sit reading absentmindedly and reach for what I take to be water. It is actually milk. But at the moment of contact with my tongue it tastes neither like water nor milk: it is a shocking and disgusting *that*—"mere sensuous matter." No wonder I may spit it out. The expectation of water has shattered, and the body-self takes a moment to realize just *what* has entered its mouth.

Shamanic Consciousness

It's not just substances like food and drink that enter our bodies. Take a wild animal. Its radiating energy and presence enters our bodies and is translated through them in some way. James says the animal itself figures in two overlapping contexts at once: as an element in the world at large, world history, and in my personal history. That numerically identical animal right there figures in the continuous central life that is myself. I am animated by it in one way or another.

James's glimpse of the intersection of body-self and the rest of the world stuns him. The first example he gives softens the shock: looking into the blue of the sky. It is more like gentle interfusion.[9] The *very same* blue that figures in world history at this place and time—this beautiful day in May on the eastern coast of the United States—also figures as event in the personal history of this occasionally conscious being, I-myself. Sometimes I deliberately look up at the blue. At other times my behavior is spontaneous, and might better be described as sky-i-fied-my-head-is-turned-up-into-the-blue—the language of possession, ecstasy, benign trance in this case.

Twelve years after his first glimpse of the shared reality of blueness, he is ready to suggest the most radical of radical empiricisms. Suggested are ancient shamanic practices of healing in which animals with pronounced regenerative powers, such as snakes and bears, figure prominently. All creativity that deeply involves the self—including self-healing—involves the labile, spontaneously involved and fusing body. If open confidently and creatively to the regenerative world, this spontaneity is deeper security. I—my ego-self—don't have the impossible task of managing everything.

But, as we have seen, the rub is that really standing open to the world can be terrifying, particularly for moderns bereft of ritual ways of mediating otherness and channeling ecstasies—particularly grievous, I think, is the loss of Mother myth. Fullness of life toward which we are pulled ecstatically, and need to be, also frightens, for we are drawn into the not-yet, the unknown, and there are no guarantees of safety. Again, the defense against this is the clenched body-self, the constricted addictive life.

James's philosophy of pure, immediate experience opens up possibilities of archaic communitarian life with a single inspired thrust of thought. It offers a ray of hope. No longer is human consciousness self-deceivingly conceived as a sealed container holding private images and sensations, the body buffering it against the shocks of the world. No longer can the screamings of injured nonhuman kin be construed as mere screechings of unlubricated machines. The very same walrus, say, that carves out its individual history irradiates and suffuses our body-selves and buoys and bolsters our personal histories—if we allow its awesome presence into us, if we do not eject it in panic. We may feel kinship with the archaic hunter who believed that the only sane way to hunt a walrus was to "shape shift" awhile and become a walrus.[10] We can identify to some extent with shamans, who, when successful, cured by entering empathically into the bodies of helpful animals and imparting their regenerative powers to clients.

We are animal organisms radically influenced by culture to form some inclusive image or sense of themselves. Reports circulate of a recent innovation in the United States called prayer therapy. On the verge of acting out a craving to eat, the client offers it up to God in prayer. Feeling oneself to be a child of God satisfies the need for significance so strongly, apparently, that the addictive trance is broken through and overridden, at least in some cases. This is not inconceivable. The current retrieval of archaic practices is remarkable, but understandable given the waywardness of the culture.

Believing we are hermetically sealed, autonomous subjects looking out at a world of objects "out there" is self-deceiving. Premature assertion of autonomy blocks its achievement in the actual world.

Neither Subject Opposed to Object nor Mind Opposed to Matter

James not only critiques subject-object thinking, but also suggests that primal body-self undercuts the distinction between mind and matter. It is neither mental—if by that we mean nonmaterial—nor is it material—if by that we mean nonmental. As we must undercut the polar opposition between subject and object, so we must also between mind and matter, and find new ways of understanding ourselves. Better to say that body-selves are sites through which energy circulates and exchanges with the rest of the world.

Taking a clue from related thoughts of Benedict Spinoza's three hundred years ago, we can say that what we call "body" is just one aspect of an energic system, another being "mind." This is somewhat analogous to a curved line that looks convex from one point of view and concave from the opposite. "Mind" will be "minding" and the whole body will be doing it, though not necessarily consciously. This understanding conforms to recent discoveries in neuroscience.[11] These suggest that one aspect of our energic system is neuropeptide activity, and the other aspect is thinkings and feelings. Christiane Northrup, M.D.:

> Not only do our physical organs contain receptor sites for the neuro-
> chemicals of thought and emotion, our organs and immune systems *can*

themselves manufacture these same chemicals. What this means is that our entire body feels and expresses emotion—all parts of us "think" and "feel." . . . This gives the person the capacity to modulate her own pain without medication. . . . *The mind is located throughout the body.*[12]

Marvelous, except I think it's better not to say that "the mind" is located anywhere. "Point it out!" the skeptic will sneer. Better to say that when the energic systems of an environment are translated in a certain way through the energic system of the organism, neuropeptide activities located in the body occur *as well as* activity we shouldn't try to locate precisely—a certain thinking and feeling. And all this will be either regenerative and constructive or degenerative. Thinking and feeling are ways the world comes to presence for us, and it's not that they are not fully real, but they are not the sort of "thing" that we should expect to be precisely locatable. Candace Pert:

> Consciousness isn't just in the head. Nor is it a question of mind over body. If one takes into account the DNA directing the dance of the peptides, [the] body is the outward manifestation of the mind.[13]

Yes, but I'd rather say the body is the precisely locatable aspect of the body-self energic system, while "the mind" is the imprecisely locatable and measureable sense or meaning that the body-self makes as the rest of the world pours through it. This lack of precise measurability and locatability is part and parcel of the privacy aspect of the self, the zone where objective standards of public consensus are inapplicable, and which must be respected if persons' autonomy is to be. More of this in subsequent chapters.

The primal body-self is "neutral" between subject and object and between mind and matter. But because "neutral" badly misleads, suggesting the innocuous, I will usually say *primal*. Indeed, the body's lability disturbs or blurs all conventional categorizations and polar distinctions such as voluntary/involuntary, particular/group, human/animal, good/bad, potential/actual, biological/cultural, even, to some extent, male/female. Loaded with unpredictable power, the body is dangerous, archaic and potentially ambiguous, has *mana*, and must be treated with the special care of the sacred or taboo.[14] Understanding how it loses itself ecstatically in the regenerative world helps grasp the sacramental and regenerative life. Understanding how it loses itself in an escapist world helps grasp the addictive and the demonic.

The Spontaneous Body-Self

Perhaps "serpentine" best describes the primal body-self, for it catches the fascination of watching a snake propel itself along the ground: we don't know where it will go next, nor how it moves at all. The snake's danger lies in its exploratory movements' uncanny power to surprise. Yogis have described this as the Goddess *kundalini*, the energy that lives coiled in the human body, at the base of the spine and in the viscera.[15] Aborigines honor the snake. Dewey writes of the uncanny instability of immediate experience, and the primal body at its tremulous core. The spontaneous body, poised in its very lability, *is* taboo—attractive-frightening, fascinating-appalling.[16]

Refusing to accept the body's primal need for spontaneous movements opens the way for frightened substitute gratifications and addictions. Recognizing these movements' powers offers hope. Spontaneity can be cultivated, its terrors transformed. We can learn from indigenous rituals in which wilderness spontaneity was celebrated by being repeatedly reincorporated in age-old rhythms of wilderness Nature. It did not run wild.

Habitual attempts to turn body into just an object, to manipulate and force its gratifications, alienates body-selves within themselves addictively. The serpentine body-self is not easily pent up and managed. Despite our being nine millennia removed from hunter-gatherers' roving existence, millions of years of prior adaptation leave a living residuum of spontaneity. It is remembered nostagically in myths formed within the time of agriculture, and in the Bronze and then the Iron Age. Paul Shepard:

> The universe, originally said in India to have come from parts of the cosmic serpent, Vrtra, was dismembered by Indra, much as Tiamat had been cut up by Marduk in the Sumerian myth. "The sacrifice of the primal serpent of the oceans is the measuring out of time, and the divisioning of space," says Jyoti Sahi.[17]

When time is measured out as linear sequence merely, the actual time we live is repressed. We don't eat when hungry or sleep when tired, but are geared to the clock.[18] In archaic time no chronology is possible. The past is a reliable, propelling presence, and we enter the future whole and confident. Likewise space is not parceled out in fixed domains that must be protected from land-hungry others. The serpentine body has room to roam and is spontaneously at home in the wide world that formed us. There is no reason for neurotic fear or addictive cravings.

We were driving on a side-road in the Outback of Australia and saw Aborigines in small groups amidst shelters and bushes. They sat on the ground, looking to me as if they were animals sitting on their haunches. That I had thought this distressed me. Later we found others who likewise retained some of their culture and themselves, who had not given up to European "spirits"—science, technology, money-profit, or more likely, alcoholism, drug addiction—and who were approachable. I learned what was happening when they squatted on the ground "like dumb animals." Without averting their eyes from me, they would say, through someone who spoke English, "See the sea eagle returning with a catch." I had seen nothing. They would point upward, again without averting their eyes. There was the eagle. And there was something hanging from its talons. They had seen it in the farther reaches of their visual and empathic field. Or they would indicate to their companions to attend to the food supply, for, without turning, they had heard the rustling and nibbling of small creatures behind them. Squatting there, they were poised to move themselves in any direction spontaneously.

They probably saw me plunging and tunneling.

The Involuntary Moment

Perhaps nobody of European ancestry better catches our immediate, spontaneous openness to the world—our birthright from ancient times—than does William James. Presenting a contrast to voluntary attention, to what requires varying degrees of effort, he writes of

> passive immediate sensorial attention . . . [to what] appeals to some one of our normal congenital impulses and has a directly exciting quality . . . we shall see how these stimuli differ from one animal to another, and what most of them are in man: strange things, moving things, wild animals, bright things, pretty things, metallic things, words, blows, blood, etc., etc., etc.[19]

Without this immediate footing in organic life, voluntarily directed attention would be impossible. Passive attention is necessary to recover from the often fatiguing stretches of the voluntary sort. It is particularly needed today given our highly specialized lives. In modern urban conditions the whole body-self is not routinely exercised, and in the effort to keep out distractions voluntary attention typically becomes compulsive or addicted and chronically fatiguing. In the well-rounded activities of hunter-gatherer forebears, voluntary and involuntary attention seem to have alternated intimately and organically; the stress of effort was, presumably, not chronically fatiguing, but ecstatic, and body-self was refreshed in the fullness of time. Since urbanites' need for recovery after stress is not adequately ritualized today (even our sleep is managed and stressed), our greatest need—to fully *be*—is not met. This lack precipitates addicted and exhausted lives for many. Involuntary attention—if allowed—can merge us with what is incomprehensibly full and great and proceeds in its own rhythms, and, of course, restores us.

Experiences in wilderness today only approximate those our hunter-gatherer ancestors knew. Yet studies with some claim to being scientific or empirical confirm the benefits of wilderness outings, the more prolonged the better.[20]

James's most alerting and enlivening example of involuntary attention, and of the spontaneity and precariousness of what he calls "the instant field of the present" (not to be confused with Descartes' mathematical "point instant") is motion: It's a "what," but it can't be applied to any one thing. It is generalized motion, as when we sit in a train and cannot tell at first whether it's our train that begins to move or the train on the track next to us seen through the window. His most trenchant example: "When clouds float by the moon, it is as if both clouds and moon and we ourselves shared in the motion."[21]

But if at this moment we cannot distinguish our motion from that of the rest of the world, a feature essential to our individuation as body-selves is missing. When such a moment occurs unexpectedly in unprotected settings, it is a threat to our being, and helps explain the primal need to be included in groups that promise ritualized ways of stabilizing and defining their members. But if the setting is more or less protected, and we are skilled at living, this momentary lack of differentiation between moving body-self and other moving things can figure in what Muir called divine frenzy—the momentary loss of balance that adds to the excitement of creativity. In this excitement we partake of wilderness, will-of-the-place. We are coherent.

Notes
  1. Jean-Paul Sartre: "One puts oneself in bad faith [self-deception] as one goes to sleep and one is in bad faith as one dreams. Once this mode of being has been realized, it is as difficult to get out of it as to wake oneself up . . ." (*Being and Nothingness,* translated by H. Barnes, New York, 1956, 58).

2. *The Concept of Dread*, 37–41, 55.
3. See Andrea Nye, *Words of Power*, and Arthur Evans, *Critique of Patriarchal Reason*.
4. James, *The Varieties of Religious Experience*, 116. Compare Bishop Bartolome de las Casas, *The Devastation of the Indies*, as he writes in the sixteenth century of destruction of indigenous New World peoples: "Our Spaniards have no more consideration for them than beasts . . . But I should not say "than beasts" for, thanks be to God, they have treated beasts with some respect; I should say instead like excrement on the public squares" (quoted in D. Maybury-Lewis, *Millennium: Tribal Wisdom and the Modern World*, New York, 1992, 16).
5. Midgley, *Beast and Man*, especially 303.
6. For mimetic engulfment, see my *Role Playing and Identity*, and references to Edward Hall.
7. *The Principles of Psychology*, I, 322.
8. There are contemporary versions of Cartesian dualism. For example, the illustrious theoretician of linguistics, Noam Chomsky, who argues in *Cartesian Linguistics* that human language is utterly distinct from other animals' signalling, and is an independent system. This may very well be true. We *are* strange! But this presumptive truth easily prompts a disastrous inference: that in actual human beings human language is independent of, and on a level above, our "mere animal signalling." No, it too often functions to disguise our fundamental message in our signalling. The two are in deepest complicity. See Sarles, *Language and Human Nature*, 87.
9. His first sketch of his philosophy of pure experience or radical empiricism appears in 1892, in the last pages of the abridgement of *The Principles of Psychology*.
10. See Calvin Luther Martin, *The Spirit of the Earth*, and Paul Shepard, *Nature and Madness*—both indispensable.
11. Better to say this than brain-science, for "the brain" by itself is no more an autonomous unit than is "the mind": exclusive scientific focusing on the brain probably reflects the furtive influence of "the mind." For a recent version of "neutral monism," see neuroscientist Karl Pribram, *Rethinking Neural Networks*, 517, 531–535, where both mental and brain states are conceived in terms of units of information. Also see the physicists Pauli (*Writings on Physics and Philosophy*, 159, 260) and Bohm (in Renée Weber, 38).
12. Northrup, *Women's Bodies, Women's Wisdom*, 29–30. Also Baars and Penrose, Sources.
13. Quoted in Northrup, 25. And see Pert's (and R. J. Weber's) "Opiatergic Modulation of the Immune System," in Muller and Genazzani, eds., *Central and Peripheral Endorphins*, 35 (listed under R. J. Weber).
14. For *mana*, see Lucien Lévy-Bruhl, *Notebooks*, and Weston La Barre, *The Ghost Dance*, 384. In general, *mana* means the presences of things that inhabit or haunt other things. It is sometimes thought to be a generalized power that runs through all things. Everything feeds out into everything else, and all feeds back into it.

15. Sri Chinmoy, *Kundalini: The Mother-Power,* Jamaica, N.Y., 1974.
16. Note Thoreau standing on the peak of Katahdin. It is time to enlarge what was earlier quoted: "I stand in awe of my body, this matter to which I am bound has become so strange to me . . . I fear bodies, I tremble to meet them. What is this Titan that has possession of me? Talk of mysteries! . . . Contact! Contact! Who are we? Where are we?" *The Maine Woods* (1864), also published in *Thoreau in the Mountains,* New York, 1982, 150, ed. by Wm. Howarth, who rightly refers to this as a climactic meditation, a turning point in Thoreau's career (152).
17. Shepard, *The Others,* 260. Sahi's book, *The Child and the Serpent,* London, 1980.
18. *Utne Reader,* "How to Tell Time Like a Cow," September–October, 1997, 80: "As my over 300 pound uncle replied when I once asked him if he was hungry, 'I haven't been hungry in 45 years.' "
19. *Principles of Psychology,* I, 416–417.
20. See chapter 6.
21. My anthology, *The Essential Writings of William James,* 205.

PART II

Addiction: Circular Power Short-Circuited

∼✖✖∽

For the good that I would I do not; but the evil which I would not, that I do. Now if I do that I would not, it is no more I that do it. . . . I see another law in my members warring against the law of my mind, and bringing me into captivity . . . O wretched man that I am! Who shall deliver me from the body of this death?

—PAUL THE APOSTLE

Activity is not, as ideology teaches, merely the purposive life of autonomous people, but also the vain commotion of their unfreedom . . . blind functioning. The subject is yoked into the world's course without finding himself reflected in it or being able to change it . . . hope . . . has deserted a subject thrown back powerlessly on his own resources.

—THEODOR ADORNO

Generations have trod, have trod, have trod;
And all is seared with trade; bleared, smeared with toil;
And wears man's smudge and shares man's smell: the soil
Is bare now, nor can foot feel, being shod.
And for all this, nature is never spent;
There lives the dearest freshness deep down things . . .

—GERARD MANLEY HOPKINS

Chapter Five

The More Than Merely Human:

Hunger to Belong

The world—this . . . *other me*—lies wide around. Its attractions are the keys which unlock my thoughts and make me acquainted with myself.

—EMERSON, "THE AMERICAN SCHOLAR"

We live in a universe . . . in which . . . all is one and at the same time all is many. The extra rooster and I were subject and object until one evening we became one.

—GARY SNYDER, *THE OLD WAYS*

St. Paul's confession in Romans chapter 7 is one of the greatest of its kind: the "law in his members" wars against "the law of his mind." Who shall deliver him from "the body of this death"?

But without careful interpretation his confession may mislead. It can prompt dualistic thinking and the equating of self with mind. The body and its organs or "members"—their urges and cravings mere brute facts or forces—may seem to lie outside mind and self. While providing a plausible-sounding explanation of addiction, this division blocks real understanding.

Without grasping that body is essential to self and mind, we can never fathom why addiction grips so fiercely and so bafflingly. Why should there be feelings of guilt if the body's urges "just made me do it"? Guilt implies some feeling of agency, and the profound guilt of addiction implies profound agency—what I-myself do.

And indeed my body-self is acting, so it is I that act. But because I cannot willingly own up to and accept the need whose addictive substitute is the bodily craving, it is also not I. Insidiously, the guilt perpetuates the entranced ambiguity and the addiction, for the addictive gratifications distract from the guilt. Take a morning-after remedy—"a little hair from the dog that bit you"—and the trance is maintained.

The dog that "bit" me—Buppie—was a very different matter. He couldn't displace his own primal gratifications with substitutes, and he was teaching me something I had nearly forgotten: we know the world with our noses, hands, feet, bellies, not just intellects. And he, a domesticated animal, led me toward wild ones like the red fox, and the wild stratum still there in myself.

In noting hunter-gatherers' spontaneous identification with wild animals, Paul Shepard recalls Dewey's admonition to emulate animal grace.[1]

I had just returned from a conference where my paper had been well received. Gratification bubbled and frothed. Now emptiness and restlessness set in. Buppie looked at me lonesomely through the gaps in the bushes surrounding his pen. I clipped the limber chain to his collar and we ran in the fields. Everything fell into place.

Caught up within a scientistic culture I had to both cling to the dog and find my way back to origins through thinking. There are any number of ways of splitting off and forgetting who we are.

Loss of the Wilderness Self

The good-enough mother—D. W. Winnicott's phrase—is so confidently connected in the world around her that her children feel safe to explore. When, in the fullness of time, they leave her, they can enjoy the wilderness of their own spontaneities because they feel competent in will-of-the-place, Mother Nature.

Paul Shepard:

> The growth of self-identity requires coming to terms with the wild and uncontrollable within. Normally the child identifies frightening feelings and ideas with specific external objects. The sensed limitations of such objects aid his attempts to control his fears. As the natural containers of these projected feelings receded with the wilderness, a lack of substitutes

may have left the child less able to cope and thus more dependent, his development impaired.[2]

With loss of competent wilderness self in the Neolithic age, fear of loss of stores had to be counteracted. More and more had to be accumulated. Consumerism began in this, the agricultural age, Shepard believes. Endemic fear took over whole societies, and autistic repetition compulsions emerged, a generalized addictiveness.

> These anxieties would elicit a certain satisfaction in repetitive and exhausting routines reminiscent of the swayings of the autistic child or the rhythmic to-and-fro of the captive bear or elephant in the zoo. Early farming represents a "state of surrender," says anthropologist Daphne Prior, and the farmer was a defeated captive.[3]

Not so long ago, my son and I climbed Mount Goddard in the Sierras. As we wound up the initial high shoulder, beginning at about 12,000 feet the other mountains fell farther beneath us, and the range opened out more and more on every side. At about 13,000 feet we stood at the center of the vast circle of the horizon: the World-whole became evident. I was elated—words poured out of me.

My son, a medical student, became impatient to get to the top. He lured me with a plastic container of chocolate milk. I proceeded as rapidly as I could. After gaining about five hundred feet elevation, I rounded a bend and my knees started to give way. The mountain had become a knife-edge a few feet wide upon which I found myself sinking down very slowly and uncertainly to my knees. The mountain dropped away on both sides for thousands of feet.

The possibility of falling was a strange actuality. Trembling, I tried to keep balanced as I collapsed, sinking fingers like claws into the rocks, trying to hang on, as near motionless and inert as I could be in that weird vacuum in which I blurred with the world. I could not pull myself safely within myself, that is, I could not connect confidently with the world around me.

Something other than rational fear occurred. An ancient alarm sounded warning that a primal need for support, unremembered in consciousness, was going to go unmet. But because the need could not be identified and satisfied at that moment, the response to the alarm was panic. Indeed, the alarm was stuck, and when not blaring out as it did that day was a source of constant unease, not focally noticed, but there.

When primal needs are not reliably met, we are ever set to be alarmed and to shoot wildly. Even when an addiction is "under control," the possibility of relapse may beckon. The distinction between actuality and possibility is not clear and sharp within prereflective immediate involvement in the world minute by minute. Possibility has its own reality, and fighting it off is a kind of meta-addiction, which may supply some continuity for a life. In this way a constricted and pressing drama may preoccupy, say, the member of AA, fixed on taking One day at a time.

Beyond the Merely Human—Encounter with the Other
Tempting to think, "This envelope of skin that encloses these bones and innards constitutes me, I-myself. Everything outside is other than me. When, for example, I touch myself there is touching and touched together. Self is this indissoluable complex of experiencing and experienced. Self is this reflexiveness, this self-consciousness. If another touches me, I have an experiencing of his or her touch experienced by me, but the other is only experienced. I am touched by the other."

Ironically, this truth about our individual being greatly misleads because, applying only to some experiences, it presumes to cover most times and situations. The primal body we live and are is not a self-sealing object *or* subject, but a resonating node of the world through which energy exchanges freely and fluidly or constrictedly and addictively.

When body-self is confidently diffused, that is, experiencing body marginal in consciousness but securely rooted, the world is this inexpressibly intimate and companionable "other me," as Emerson put it, with no clear line dividing self's and other's experiencing.[4] We vibrate and resonate with others' feelings and thoughts, though we may not know exactly what they are thinking just as they are thinking it. States of body-self vary from this total, or almost total, joyous merging, through a continuum of states to the body-self's frightened attempt to close itself up within the keep of its body's surfaces. If the "other me" violates the body-self's frightened closure, it shifts its signature with terrifying abruptness: at worst, one is demonically possessed, as we will see. Chronic fear of this prompts addictive repetitions by body-self, essentially hysterical attempts to reassure itself of its integrity through time.

There are many moments of anyone's experiencing in which self-consciousness—that is, focal awareness of our clearly demarcated bodily envelope—is practically nonexistent. We are caught up intimately in the surround—the rooster and we "become one," as Gary

Snyder writes. Then my experiencing is not sealed off from others' experiencing. Even "inanimate objects" (so facile our language!) are more like fellow subjects, intimate participants in one abounding life. So the neat formula, "I-myself am both experiencing-and-experienced while others are only experienced," applies only to moments of defensive, smug, or cool self-containment.

At the instant Gary Snyder hears the rooster's crowing song, he experiences blissful union with the bird. If he had perceived a threat to himself, he would have reaffirmed the line dividing body-self and the rest of Nature. But why forget ecstatic and numinous moments of union? Typically they alternate in the blink of an eye with narrowly alerted ones in which the unique value and distinctness of located and rooted body-self are reaffirmed. But experiences of union are profoundly real, and they shape our being, as do the other moments. Without them we lack the momentum to pass by the whispered temptations of immediate gratifications, addictions.

The connection between self and other is extraordinarily elusive, subtle, complex, varying and various. For example, it is not always true that the more alone we are, the more certain we are of our individuality. When abandoned we may anxiously feel our reality slipping away, sublimating into the void. Ordinary language is apt: we are beside ourselves. The steady touch of the reassuring other can enter and center us, restoring us to ourselves.[5]

Perhaps it is not a human other that touches and restores me. A round stone cupped in the hand might do. Or a rosary or set of "worry beads" fingered in sequence. Or a soft breeze, coming from far off to find me, that rustles the hair on my arms. Without the reality of the sustaining universe with which we identify, we would be less than ourselves. Without feeling its sustaining power moment by moment we fall into black holes in time, panics large or small.

The World As the Greater Body

We took shape over many millennia within the intimate otherness of the world. David Abram:

> Our bodies have formed themselves in delicate reciprocity with the manifold textures, sounds, and shapes of an animate earth—our eyes have evolved in subtle interaction with other eyes, as our ears are attuned by their very structure to the howling of wolves and honking of geese.

But he goes on:

I found myself now observing the heron from outside its world, noting with interest its careful, high-stepping walk and the sudden dart of its beak into the water, but no longer feeling its tensed yet poised alertness with my own muscles.[6]

When Abram's momentary detachment is constant and endemic in a culture, there is vast trouble.

The reason for traditional peoples' seasonal calendar of feast days and holy days is not far to seek. Alternating stretches of prosaic involvement in everyday particulars with times of trancelike, suffused involvement in the vast and mysterious Unit, the World-whole, creates a rhythm that is profoundly regenerative. Moving from the many to the One, the One to the many, again and again, flowingly, the body-self recuperates its past, rings its changes, and stretches to its limits at the moment.[7]

The heavy black flatbed truck with three huge spools of paper standing on end and chained in place—giants' toys—turned into 43rd Street and lumbered toward the New York Times *printing presses. Electronically transmitted data were to mate with the older technology on sheets of paper. How many trees had been sacrificed? But I glimpsed that the destruction in Nature attending on that amount of paper appearing daily at this Manhattan block, involved much more than trees. Also lost would be the life-supporting fungi in the earth that feed the roots that feed the trees that support animals and birds that feed the fungi with their droppings. And then the fungi-enriched soil feeds the roots that clutch, hold, and save the soil . . . vast circuits and cycles of interdependency. Thoughts too unsettling to be held for more than a moment swirled back into the darkness from whence they had come unbidden. Lost again in the general commotion, I hurried to my appointment on 44th Street.*

Alienation from Earth and Body-Self

Some ecologists note that the irreversible wasting of topsoil, the poisoning of air, water, and remaining soil, and the expunging of untold species of plants, animals, fish is bringing the Cenozoic age to a close. This period covers the last sixty-five million years, which saw the evolution in wilderness of the marvelous array of mammals and birds, of grasses, shrubs, and flowering plants.[8] If these ecologists are right, as seems to be the case, we are destroying what humanity has heretofore experienced as the basis of life, growth, and regeneration itself.

The nineteenth-century "romantic" conception of reality as organically interdependent is just beginning to percolate into the popular (and academic) consciousness. This consciousness has hitherto been constricted by tunnel vision, that is, by addictive absorption in sterile objectivism (see chapter 10 on truth), or in egoistic and mentalistic short-range concerns; or in various spiritualistic or soulish or electronic evasions of the bodily reality of our lives. All these preoccupations are addictive, for they cannot satisfy the body-self's many needs and hungers for long, and so must be incessantly recommenced. They contain an element of the "not me," which allows responsibility for them to be displaced. Try this displacement: "Rush and busy-ness is just Americana."

But today John Muir's words in *My First Summer in the Sierra* command ever greater attention, "When we try to pick out anything by itself, we find it hitched to everything else in the universe."[9] Fortunately, biologists and ecologists lend the authority of science to Muir's observations of the astonishing scope and intricacy of Nature's interdependency, and the ways we interweave through it.

Forester Chris Maser explains why forests are not the easily renewable resource many think they are. As Dolores LaChapelle summarizes his work,

> It turns out that every tree needs root fungus to help it absorb nutrients from the soil. On the surface this fungus puts forth the mushrooms we eat. One of the things which is completely destroyed in a cut-over forest is the root fungus connection and it can't come back if you've also destroyed the northern flying squirrel or the mole which spreads the fungus for us. Maser writes: "A fecal pellet is more than a package of waste products; it is a 'pill of symbiosis' dispensed through the forest . . . it inoculates the soil with live fungal spores, nitrogen fixing bacteria, and yeast." And the squirrel does it all for free![10]

Deeply ingrained and now exacerbated tendencies to conceive of self as other than body, and aversion to the body and its products, easily involve inattention to fecal pellets. Aversion to body generalizes itself unconsciously as aversion to the intricate interdependencies of all of material Nature that support our lives.

Immensely important as are strictly physiological and biological analyses of Nature, they cannot be sufficient. We sense what else is required when Maser speaks of the forest, not as something merely "out there," but as our living body without which we would die. Our corporate body includes more than the corporate body of human groups. It includes the whole living world—at least that. And talk about body in its relational amplitude and lability is talk about self.

The forest as our living body without which we would die? Mere metaphor? Or a little magic slipped in? But we either make magical thought and myth our own, or some distorted fragments of them run away with themselves and make us their own. Electronic technology has quickly become a weird spiritualism breeding science-fiction fantasies that detach us from the body we have become over countless millennia, and from the body of the Earth that has always molded us rhythmically within itself.

Finding Our Way Back

In the ancient world, myth-making peoples were confirmed and built their selves through communion with the natural world. As Chellis Glendinning puts it deftly, they enjoyed

> a kind of natural psychotherapeutic process that involves projecting our grapplings with growth, conflict, and triumph onto natural beings, like bears and lakes, and working them to inner resolution by imagining, by telling stories, and making rituals.[11]

This vicarious belongingness is a real magic. We are always in communion with the intimate otherness of the world, but the primitive wisdom affirming this truth is easily lost. Without it we perpetually try to objectify the world and so become stranded subjective selves, or half-selves. Glendinning writes of loss of healing reciprocity:

> And what follows losses like these? Often, it is the addictive process: a reaction to a loss of satisfaction of our primary needs. It's a rather ingenious attempt by the organism to satisfy primary needs with secondary sources.

When seventeen or eighteen I was fishing one summer's day in the New River of northern California. Feeding down from its steep canyon, the New enters the gorge of the Trinity River in a section that must have been difficult for white settlers to find. That is how the river got its name: it was new to them. Augmented by this tributary, the Trinity flows some forty miles to the stately Klamath, and the Klamath flows thence to the sea.

The New is either a small river or a very large stream, averaging about twenty-five feet across. But many of its pools are seven or eight feet deep, and in them salmon as large as twenty-five or thirty pounds swim lazily and rest after surging up through rapids or falls. They neared their destination, their origin, where they would spawn and die and propagate their young, just as they themselves had been given the gift of life by their dying ancestors years before.

Fish that large in a river so small—that made it a special place. On this day I was fishing alone in the canyon of the New with its bluish granite, grand fir trees, and the great fish finning so near in the crystal water. In awe at their numinous presence, I fished in the most desultory way. Moving slowly downstream, I reached a point where the granite walls closed in on the river and prevented my going further.

I sat on a smoothed hummock of granite and watched the fish. I could proceed no further . . . unless I swam. I felt that I did and did not belong there in the water with them. I was attracted/repelled, fascinated/fearful, pleasured/ pained, the twin emotions in tremulous balance. I found myself taking off my boots, then socks, then pants and shirt. I threw myself into the water, and without daring to put my face under, swam the eight or nine strokes to the rocks on the other side. I pushed myself up on them and sat shivering. It was perhaps the most important thing I had ever done.

I began to relax more deeply than at any time in my life. In those moments in the water I had not been merely the fishes' killer. Dimly and dumbly, but really, they must have recognized me as a fellow living thing. Thus recognized, I was more fully a living being, more fully myself. My larger countenance reflected back and shown within me.

It has taken a long time to understand this.

Body-Self and Native Peoples

When we urbanites contrast ourselves with an indigenous population of this continent, the Yurok nation, we better understand ourselves. The Yuroks' world, and their rooted, mythic integration with it, gravitates around the Klamath River on the coast in northern California. This is brilliantly described by Freeman House in his "Totem Salmon,"

> Salmon is the totem animal of the North Pacific Range. Only salmon, as species, informs us humans, as a species, of the vastness and unity of the North Pacific Ocean and its rim. The buried memories of our ancient human migrations, the weak abstractions of our geographies, our struggles toward a science of biology do nothing to inform us of the power and benevolence of our place. Totemism is a method of perceiving power, goodness, and mutuality in *locale* through the recognition of and respect for the vitality, spirit, and interdependence of other species. In the case of the North Pacific Rim, no other species informs us so well as the

salmon, whose migrations define the boundaries of the range which supports us all. . . . It is the nature of industrial capital that it has no interest
in preserving indigenous populations, since capital is mobile and can
move on once a "resource" has been exhausted. Indigenous populations
cannot flee this rapacity because of their biological marriage to habitat.
What is not generally recognized is that the human species is also an
indigenous population. We too are inextricably married to place. We can
only be kept constantly informed of our situation as a species through
regard and recognition of brethren species. The life of the wild salmon
population is of the essence of the life of the human population.[12]

It is unreasonable to expect that animals, fish, and birds can orient us
as totally and precisely as the salmon do the Yuroks. We have modified
our relation to Nature through technology too drastically. But without
resonating occasionally with wild regenerating Nature, something in us
will die.

Finding a life shared with salmon: nothing had taught me to speak
of such a thing. It filled me with incredible life, at least for a while, and
before I forgot it.

My wife slid the current copy of Newsweek *across the kitchen table and said
jocularly, "Watch a romantic get zapped." The magazine was open to a con-
temptuous review of* The End of Nature. *It was sub-headlined, "Bill McKibben's mournful reflection on the environment is based on a romantic fantasy."
The reviewer (Jeffrey Cowley) continued archly, "when [McKibben] walks in the
woods these days, the rain lacks that old 'independent and mysterious existence,'
and the wind no longer comes from 'some other sphere.' It never really did, of
course, wind, rain and human life are all natural phenomena."*

*Sensing that it might be an important book, I immediately bought a copy. It
is. Very simply, on a gut level, without any explicit allusion to boundaries or
horizons, McKibben relates how his knowledge of the alteration of the atmosphere through pollution has altered his experience of Nature. Indeed, no longer
is Nature experienced as independent of us, as "coming from another sphere."
In other words, he no longer feels that vivifying contact across the boundary into
primal wilderness, what I would call the ever-ineliminable but ignorable matrix
of our life.*

*The reviewer exemplifies perfectly our culture's dominant secularism and sci-
entism. According to this writer no "natural phenomenon" could be mysterious
or sacred—by definition it could not be. According to him, "this isn't a book
about solving real-world problems. McKibben's ultimate concern seems to be his
own state of mind." Again, the review exemplifies the dominant view: attitudes*

such as McKibben's can only be private states of mind, not real matters of concern.

But where are we left when our attitudes aren't regarded as real matters of concern? When, particularly, any sense of the sacred or of mystery is dismissed as mere personal idiosyncrasy? Unless we are very mature, we will feel like zombies, and as we feel, so we tend to be. *People feeling empty and not fully real, feeling like zombies—that* is *"a real-world problem" for our society.*

The reviewer breathes contempt and alienation so complete that he cannot imagine himself to be alienated—closure, the whiff of stagnation from the bottled-up managerial mind—conventional wisdom at its ridiculous worst.

Body-Self Opening up New Consciousness

Our deepest problem is our inability to imagine the magnitude of what we are doing to the Earth and to ourselves. We need new modes of world-consciousness and a visceral sense of being at home in Nature. But how to attain these? To be apprised of some of the gross facts of environmental degradation helps some. But I would also return to the lessons of Thoreau, Emerson, James, Dewey, and the shamanic consciousness they direct us European-Americans toward.

The scope of our identifications and sympathies can be broadened, I believe. If we are somehow both bold and relaxed we may generate a new consciousness.

Walt Whitman: "I sing the body electric."

I was discharged from the army in 1955 and went to live in New York City. Sometimes on Sundays I'd go to the nearly deserted Wall Street area in downtown Manhattan. For me this would be "going to the mountains." It was the great buildings, the narrow canyons between them, and the silence that I liked.

Father had taken me to the High Sierras when I was four or five. His daily reading at home of the stock market reports, those long columns of mysterious, minute figures, must also have been evoked. In any case, as I walked one day in downtown Manhattan—this time on a weekday with its crush and bustle—I felt as if my father's body had entered mine. His peculiar hitched-up walk was in my pelvis and legs. Unmistakable, he—his presence or whatever—moved in me, moving me. I was speechless, fascinated, elated. Again I experienced how much I loved him.

I thought at the time of telling others, but did not.

Notes

1. Shepard, *Nature and Madness*, 37–38. Dewey, *Art as Experience*, 19 & 25.
2. *Nature and Madness*, 33. Adolescents' attempts to frighten themselves with

horror movies may be a covert attempt to reassure themselves: "You see, we made it, we still live." But this may be counteracted by the realization that "the horrors weren't real." Movie technology confuses.

3. *Nature and Madness*, 34 & 43.
4. See Erwin W. Straus, "The Expression of Thinking," in *An Invitation to Phenomenology*, ed. J. Edie, Chicago, 1965, 266 ff.
5. See Catherine Keller, *From a Broken Web*.
6. "The Ecology of Magic," *Orion*, Summer 1991, 38–39. See also Abram's more recent *The Spell of the Sensuous*.
7. For the principle of alternation, W. E. Hocking, *The Self: Its Body and Freedom*, 404 ff.
8. Note this Forest Service sign, Seneca Rocks, Monongahela National Forest, West Virginia: "Lichens are . . . two plant species living very closely together. The fungi part provides the anchor while the algae provides food. Some lichens can attach to bare rock and can actually dissolve it, and they are considered pioneer plants and prepare the way for other plants. . . . some species can . . . live for 4500 years . . . soil building is a very slow process." Can we begin to understand this on a level that will actually curb our fantastically out-of-touch wastage of topsoil?
9. Muir, *My First Summer in the Sierra*, 81.
10. Dolores La Chapelle, *Eighth Annual Newsletter*, Way of the Mountain Learning Center, Kivaki Press, 585 E. 31st St., Durango, CO, 81301. Maser's book is *The Redesigned Forest*.
11. "Dreaming for the Earth."
12. Obtainable from Way of the Mountain Learning Center, see note above. It is also forthcoming from Beacon Press. Concerning precipitous decline in runs of salmon, *The New York Times*, front page, March 7, 1994.

Chapter Six

Medical Materialism and the

Fragmented Grasp of Addiction

*In college I tore ligaments in my knee playing intramural football. Since I could
no longer kick, I passed the ball more frequently and strenuously than ever.
After some years I tore the rotator cuff in my shoulder. I again took up kicking,
and seventeen years after the initial accident found I could no longer straighten
my leg. Early on the morning of the operation I was awakened, feeling nau-
seous, inert, and apprehensive. An injection was administered. Very soon I was
happier than I had ever been. As I was wheeled on my back on a gurney toward
the operating room, the ceiling flowed past and I felt as if I were a dolphin
leaping out of the water. Incredible euphoria. The last thought before losing
consciousness was, Maybe they will operate again soon so that I will get the
syringe again.*

The conventional medical model of addiction emphasizes natural sciences of the body—biology, physiology, biochemistry, genetics, and so
on. These are necessary for understanding addiction but are insufficient. For in order to do its important work of precisely correlating
processes and trying to establish causal linkages, science must objectify
the body and quantify it. In so doing, however, it masks out crucial
features and qualities of body as self.

Body-self's immediate experiencing and minding of the world cannot be objectified or quantified by anyone, not even the person himself
or herself. But it is only in this experiencing that the addictive condition really shows itself: the insistent presence of substitute urges and
constricting satisfactions that cannot be made fully one's own, that cannot satisfy for long, and which must be repeated on pain of suffering
withdrawal.

Likewise, it is only in this immediate experiencing that nonaddictive, integrated, ecstatic involvement of the body-self in the World-whole shows itself. This may be spontaneous and joyous integration. Or it may be integration accomplished only by gritting one's teeth and enduring when difficulties and tragedies occur. Thus character shows itself, along with that ground-mood of assurance in the midst of tribulation and temptation essential to character.

Because science must restrict its focus in order to do its work, it can produce only a partial account of addiction. This account will be deceptive, however, only if it is accompanied by the scientistic dogma that only science can know.

Addictions As "States of Physical Dependency"

It is understandable that certain addictions are described by natural scientists and physicians as "states of physical dependency." The body is thrown into a precarious and ultimately harmful new equilibrium that can be maintained only by repeated ingestions of drugs, say. Certain ones always produce what physicians call physical dependency: "an adaptation of cell metabolism to that substance," such that termination produces "withdrawal symptoms" (e.g., depression, tremblings, convulsions).[1]

But the effort to define addiction exclusively in "strictly objective" physiological or pharmacological terms soon comes to grief. First, ingestion of marijuana—at least for some addictive and degenerative over the long term—entails no withdrawal symptoms specifiable precisely in physiological or pharmacological terms.[2] Second, babies born to addicted mothers typically do exhibit physical dependency and withdrawal symptoms precisely specifiable physiologically. But careful observers don't describe the infants as addicts, for they have shown no will or agency in bringing about or maintaining their condition.

Following the ingrained Cartesian tradition of psycho-physical dualism, many physicians and physiologists resort to "psychological dependency" to fill out their concept of addiction. Thus in 1987 the American Medical Association formulated "chemical dependency": "a chronic, progressive disease characterized by significant impairment that is directly associated with persistent and excessive use of a psychoactive substance." But "persistent" and "excessive" were not adequately defined, not even "use" was, and the AMA granted that "impairment may involve physiological, psychological, or social dysfunction."[3]

Howard Milhorn, M.D., wrote in 1990, "There is a growing consensus that chemical dependence is a disease, because it is a pathological state with characteristic signs and symptoms and a predictable course and outcome if untreated."[4] For physicians to regard addiction as a disease gives them a handle on a desperate problem. And it is doubtless true that people who have attempted to control their lives through self-medication of addictions typically prove incapable of releasing themselves from them. They need help. Moreover, regarding addictions as a disease reduces guilt, and it is just guilt over addictions that, ironically, helps keep them going, for addictions numb and distract the addict from the guilt they generate.

Nevertheless, a great price is paid for treating addictions as a disease. Conscience is the sense of right and wrong and the feeling of responsibility. It is softened, but also the overall sense of self is reduced—the immediate sense of oneself as an ongoing source of initiatives in the world, a real power, an agent. Reduced or lost is meaningful ecstatic involvements, and the sense of responsibility that goes with them. Even if addicts receive massive help from therapists, they must take initiatives in their recovery beyond what is typically considered necessary for maladies that are normally called diseases. Milhorn speaks of addictions as "irresistible," because the hypothalamus or primitive brain with its pleasure centers overwhelms the cerebral cortex, the "seat" of agency and decision.

At an advanced stage it might be true that addictions are irresistable. (But how could we know until individuals confront, or fail to confront, their capacities, needs, and difficulties?) Yet at an earlier stage the impulse to take the drug may clearly have been resistible. And we have responsibility even if all that we can do in stray moments of lucidity is to cry for help. The danger of regarding addiction as a disease is to place it roughly on a par with tuberculosis or malaria, conditions that just happen to us.

Addiction a Disease?
To regard addiction as a disease is well-intentioned, but it is a de facto insult to human beings. It is symptomatic of the corporately addicted consumer society which fails to acknowledge the total capacities of body-selves who need responsibility, agency, and meaning in life, and who long to circulate back and forth across the boundary between wilderness and civilization. The addicted society underrates the value of wilderness and our wilderness bodies, and our ecstatic responsibility for mediating and accommodating these energies. Humanly and spiritu-

ally impoverished, we find it easy to blame addictions on chemicals that somehow get into us.[5]

But the addictions cannot be neatly confined, quarantined, within "the drugs themselves." Because before the drug was first ingested there was need for ecstasy, and it was expected that the drug would supply it.

And clearly, some behaviors must be called addictive that involve no ingestions of chemical substances. They may be called "psychological addictions," but the craving can be just as "irresistible" (a common word in the medical literature), and may drive toward goals that are just as "irrational" and "impairing" (other common words). Consider workaholism: persistent activities that when terminated leave feelings of vacuity, flatness, meaninglessness. But they are recommenced, no matter how exhausted one is. These behaviors need involve no specific physiological imbalance as their correlate or "basis."[6] Instead of thinking we have explained anything with the label "psychological addictions," we might peel it off and look behind it into the unknown.

Chuang Tzu:

> Once a man receives this fixed bodily form, he holds on to it, waiting for the end. Sometimes clashing with things, sometimes bending before them, he runs his course like a galloping steed, and nothing can stop him. Is he not pathetic? Sweating and laboring to the end of his days and never knowing where to look for rest—can you help pitying him? I'm not dead yet! he says, but what good is that?[7]

The Withdrawal Symptom Criterion of Addiction

If "withdrawal symptom" is taken broadly enough to include what some call "psychological withdrawal," it could perhaps serve as a rough and ready beginning in picking out behaviors that are addictive. But it must be greatly supplemented if we would learn much about addiction itself. For this criterion groups together obviously destructive addictions with ones that are either inconsequential or possibly even beneficial. Take repetitious coffee drinking. If terminated abruptly, withdrawal symptoms occur, so by that criterion it is an addiction. But what if, as numbers of coffee drinkers maintain, it keeps them alert, produces no serious deleterious consequences, and is a necessary ingredient in a happy and productive life? Coffee drinking is addictive, but its beneficial consequences may outweigh negative ones.[8]

But more importantly, the withdrawal symptom criterion includes within the class of addictive behaviors those that should not be called this at all. Take ardent mountain climbing or an all-absorbing artistic career or serious playing of any sort. Very difficult it is to give these up, and if they were to be withdrawn, distress would occur. The withdrawal criterion fails to distinguish activities that are regenerative from ones that are not. All-absorbing, they may preclude other modes of ecstatic behavior, but their rewards overshadow this limitation, and a life is kept vital and alive—not perfect, but vital and alive.

These I call *magnificent obsessions*. They are kinds of sacramental experience. There is skill, but abandonment in it. Voluntary and involuntary attention fuse marvelously. From the unfathomable World-whole more flows back into body-self than flowed out of it. We are fed.

When people climb mountains, risking their lives, they not only scale—measure—the mountain, they measure and confirm themselves: "I am the one who does this!" The magnificent obsession pretty well determines the contours of their identity. And they go beyond the calculable. We cannot calculate to the point that a blinding blizzard, or the merest misstep on a beautiful day, will not take our life even in the very best months for climbing Everest.

If success is achieved, the incalculable Whole flows into us, complementing beyond reckoning the greatest individual effort. The ultimate primal need is satisfied: to *be* fully, that is, to be ecstatically full beyond reckoning. If one's attempts to achieve are precluded, massive "withdrawal" symptoms occur. I doubt, however, that something equivalent to detox would be required to keep one's life going. People magnificently obsessed typically find other means to regenerate their lives. Their obsession was regenerative, not just addictive: they must have woven themselves somehow into the reality of the regenerating world. Able to test reality, they are free: capable of finding other regenerative ecstasies.

Magnificent obsessions are blessings, associated often with great achievements. There are, to be sure, disturbing borderline cases. What if the person cannot stop the behavior even if he or she sees that there are good reasons for doing so? Given this dislocation within the self, is the behavior regenerative? I doubt it, but am not sure. Studies would have to be made in each case. Perhaps we can say that a behavior is a magnificent obsession with a strong addictive strand? But what have we said? The behavior appears on its face to be unclassifiable, and reveals the great variety and strangeness of human beings, as well as the limits of our concepts for understanding ourselves.

The Addict's Brain and World

Dr. Milhorn writes that chemical dependence is "a biopsychosocial disease." But carpeting together names cannot substitute for the descriptive, conceptual, and perhaps very personal work necessary to understand addiction. Clearly, it strains all dualistic and linear categories to bursting.

Note the excellent book by the physiologist Charles Levinthal, *Messengers of Paradise: Opiates and the Brain*. He makes a concerted attempt to escape reductionism and to see life whole. But he leaves some serious questions hanging.

Helpfully, he reviews the recent discoveries of the brain's natural capacities to generate its own opiates (opioids) and analgesics—the endorphins, for example—and demonstrates how drug addictions in particular short-circuit this natural production, reducing us to dependency.

> Long ago, some unknown adventurer tasted the juice of the poppy and the brain recognized it as a creature of itself. By some quirk of nature, that plant yielded a substance that corresponded to something already in the brain. We now can start to understand why opium, morphine . . . are addictive. According to a widely cited theory, the brain essentially gets fooled. The chemicals are mistaken for what the brain has produced on its own. As a consequence, the brain stops its internal production and begins to rely upon the external source. As more from the outside is required, a vicious cycle is created.[9]

If ingestion of drugs stops, the endorphinless brain triggers withdrawal symptoms. To prevent these, drugs must be ingested continually, and the person is caught in the toils of addiction.

Levinthal then asks,

> Why, then, do we not become addicted to ourselves? . . . The easiest way of answering is to assume that the quantities in which these endorphins normally operate are worlds apart from the quantities of morphine or heroin that arrive . . . from the outside. If this hypothesis proves to be true, the next task will be to develop a drug that operates at a more nearly identical level of functioning to the endorphins already in the brain. If this can be achieved, the dream of the nonaddicting pain-killer may be realized at last.

Now, how are we to tell if we have found a drug that "operates at a more nearly identical level of functioning to the endorphins already in the brain"? Levinthal would not think it sufficient to look within the brain, but rather at the whole human situation. To his credit, he even considers religious ecstasy.

But to answer such questions we must describe ecstasy much more fully still—and more, must unhitch from the exclusively objectifying viewpoint of all current science and describe the primal body-self's first-person experiencing. What if we found the "perfect pain-killer" and when we used it our lives lost some zest, some vital inflation and bonding with the world around us? Only by looking beyond the brain and understanding our direct experiencing of the periodic, regenerating world will we understand what "perfect" might mean.

The age of science engenders an ideology of scientism that is not scientific at all. It tilts into fallacious inferences such as this one: Since a necessary condition for euphoric experience is an opioid chemical in the brain, therefore *the* cause of the euphoria is the chemical. And so, the disastrous inference continues, what the euphoria is *over* or *about*— its meaning—is insignificant.

But what I am happy about makes all the difference. If I am euphoric over my constructive, generous, loving, or reverential acts, the euphoria will be regenerative. If not, degenerative. What if chemical or surgical alteration of my nervous system caused a spasm of delight each time I picked up from the ground and stored any object between one inch and a half-inch in diameter or length? I would be cursed by the euphoria, and would vaguely realize this, unless I were insane.

The brain is but one element in the cycling organism-environment biocultural circuit. Regenerative euphoria depends upon the meaning and value of the whole circuit.

Meaning demands engagement with the world, with what is believed unquestioningly to be real. When a muddled fear of pain, disappointment, suffering prompts heavy use of analgesics, the shape and excitement of reality is compromised. A pain beyond pain is felt, one with fuzzy definition, fuzziness that invades one's life. Thoreau:

> Be it for life or death, we crave only reality. If we are really dying, let us
> hear the rattle in our throats and feel the cold in the extremities . . .[10]

"The Craving Brain"

A book of this title appeared in 1997 and commands attention. Written by Ronald Ruden, M.D., Ph.D., and subtitled "The Biobalance Approach to Controlling Addictions," it makes plausible claims for curtailing addictive cravings through chemical manipulation of neuro-transmitters. Not the least of its virtues is its sixty-page summary of the scientific studies supporting these claims, a majority of them published in the last ten years. The very brilliance of the discoveries in neurosci-ence, however, partially eclipses the whole environmental context within which we live, and which is essential to our sense of reality and to who we are.

His discoveries are consonant with a view of self as body-self; they could form part of such a view. Ruden contends that addictive cravings are caused by an excess of of the neurotransmitter dopamine relative to serotonin. Dopamine drives toward activity ("Gotta have it!"), while serotonin causes the feeling of satisfaction ("Got it!"). The strongest need is to survive, says Ruden, and this is inherited from creatures who lived hundreds of millions of years ago, creatures unaware of their own existence.[11] When a survival need is unmet the dopamine level surges upward and we are driven to activity. This is as it had to have been for our species to survive. If and when the need is met, serotonin is gener-ated and the activity ceases. Again, the period of restoration had to occur for species survival.

But the neurotransmitters, dopamine and serotonin, can go out of balance, and this is the root, he believes, of addictive cravings. If certain primal needs go habitually unmet, and yet the organism somehow sur-vives, a period of inescapable stress occurs in which dopamine rises unchecked and the organism clutches at anything even remotely con-nected with real satisfaction of survival needs. As I have put it, we shoot wildly. For example, if infants are not reliably fed, typically by the mother, they tend to fixate on factors associated with satisfaction, but that cannot themselves be truly satisfying and calming. If they survive at all, they "adapt," but at great long range-cost to their regenerative powers. An oral obsession and addiction may emerge, perhaps as to-bacco smoking (with, of course, its added kick of nicotine); or perhaps an addiction to ever repeated, ephemerally satisfying, sexual acts; or, again, a fetishistic preoccupation with women's shoes or clothes—a woeful constriction of one's horizons.

Ruden maintains that anything that raises the serotonin levels and balances the unchecked dopamine will eliminate addictive cravings. When Alcoholics Anonymous is successful it is so, he thinks, because it satisfies an ancient survival need: to find shelter and protection through herding. When the person is surrounded by like-minded others the need is satisfied, and the craving is abolished because serotonin balances dopamine.

But other addictions may take over because serotonin may suffice to balance only the dopamine-driven craving for a certain chemical, for example, alcohol. Ruden maintains that he is able in most cases to so delicately balance the neurotransmitters with regularly applied medications that persons are freed of all addictive cravings. He can modify "the inherent terrain of the brain."[12]

No doubt, this is an accomplishment. Yet he himself admits that much further research is needed.[13] In the too-short last chapter he asks an obvious, very troubling question: Why is there so much imbalance in brains in the first place? He briefly points out how environmental problems such as wretched poverty and ugliness can cause the inescapable stress that triggers run-away dopamine, hence addictions.

He does not grasp, however, the magnitude of the conceptual difficulties impeding research. He does not grasp what Dewey saw: that preoccupation with the brain can perversely reenact the Cartesian belief in isolated individual minds.[14] It is surely significant that he has nothing to say about what Ann Wilson Schaef describes as the addicted society—a palpable enough reality, I think. We somehow must grasp the whole reality: schematically put, brain-in-body-in-environment.

Ruden does say that pattern recognition by the brain of whole situations that cause anxiety, hopelessness, despair is not well understood.[15] But what will be needed for better understanding? Just more narrow fact-finding of the brain's neuro-chemical-electrical circuits? Ruden inadequately illuminates the *sort* of further research needed. I think we must redraw the conceptual map of the human condition. We must take James's pure experience seriously: the world experienced irradiates and engulfs the organism. Tracing linear causation between relatively isolated elements or factors has its limits. The disturbed brain is engulfed immediately in a disturbed society and Earth. We need new models of causation. The whole situation must be reconceived so that our ecstatic hunger for reality can be grasped.

Ruden's preoccupation with the brain prevents him from pursuing one of his own clues. He contrasts satisfying a primal survival need with satisfying an addictive craving. The former occasions a rise of serotonin that balances the dopamine and stills and restores the organism. But satisfying an addictive craving does not, "since there is no sensing system in place."[16] The implications of this must be drawn out. When the sensing organism is oriented to satisfying survival needs it generates sufficient serotonin because it is being true to itself and its own deepest needs: it is realizing itself in the world that formed and maintains it.

It is certainly strange that a theory so reliant upon ancient evolutionary developments of the organism in Nature should have so little to say about Nature in our conduct of life today! Ruden mentions in passing the probable benefits of "a Nature walk." This is pitifully inadequate. So much attention is given to "the terrain of the brain" that the abiding terrains of the Earth and its horizons are eclipsed. Here we must talk about the literal terrain of Earth, and about our organisms moving upon it adaptively, within its beings and its atmosphere. Whenever necessary the mature self knows, say, "I turn and that other remains stationary. That is the other, I am myself!" Belief is the feeling, the conviction, of reality.[17] The serotonin is raised and the organism quieted because we adapt with our whole body-selves to reality. We can't talk about satisfying survival needs without talking about satisfying the need for the conviction of reality.

Ruden declares "Poverty is like quicksand. It can swallow up one's sense of self."[18] But "sense of self" cannot be grasped in neurotransmitter terms alone. It requires those that must somehow be connected to scientific ones: freedom, accomplishment, responsibility, reality, conviction, love, hate, etc. And, indeed, character: the "dead heave of the will" (James's phrase) which may be required to face the possibility, then the reality, of one's addiction, and to seek help—perhaps from Ronald Ruden himself.

Moreover, without in any way minimizing Ruden's and others' scientific achievements, we wish they would tell us more about the lives of those who have been freed from addictive cravings. Does the freedom last? What is the ecstatic quality of their lives? What is the long-term effect on body-selves of dependency on the medications that balance neurotransmitters? Wouldn't it be better if persons could be weaned from these eventually, so they might be free to synchronize their behaviors with the regenerative cycles of Nature in which we were formed—the natural cycles of stress and restoration?

Ruden's work epitomizes the mechanist and reductionist argument that consciousness is nothing but an aspect of the brain because, the argument goes, if you put drugs into the bloodstream you alter the function of the brain and the state of consciousness. This is true, but it does not follow that either brain or consciousness functions in isolation from the larger world. To cut through the organism-environment circuit at the point of the brain and to think that what's discovered at that cross-section is *the* answer to what's going on, is fallacious reasoning. Ironically, the stunning truths the scientist discovers prompts the fallacious inference.[19]

In sum, a condition for the advance of science is its restricted vocabulary and sharp focus. But we make sense of our lives and fully *be* only because we make sense of wide-ranging evaluative and normative experiences and ideas—for example: *fully being, achievement, valuable, civil, regenerative ecstasy, holy* or *sacred*. Natural science cannot deal with these, and we shouldn't expect it to do so. Expecting it is scientism, not science.

Designer Drugs

Until the late 1980s the use of drugs to alter "the mind" had been limited to obvious pathologies and to the psychedelic trips of those pulled into the vortex left by the 1960s. Spectacular very recent discoveries concerning neurotransmitters, coupled with the tide of technological, managerial, consumer culture of the 1990s, have radically expanded the use of psychoactive drugs in general. They bode to become as socially acceptable as hair dye, megavitamins, contact lenses, or plastic surgery. But basic issues are not resolved, and instead are highlighted and exacerbated: What is the real self? What is regenerative ecstasy?

Within a decade, probably, a large segment of the population will be taking some sort of psychotropic neurotransmitter-manipulating drug.[20] Many believe that a distinction must be made between legitimately prescribed antidepressant and antianxiety drugs, on the one hand, and the drugs of addiction on the other. I have noted that the concept of addiction is fuzzy on its edges, with some behaviors borderline and undecidable. But if the question, are the drugs the psychiatrist prescribes addictive, is to be decided in a significant number of cases, the criterion must be the regenerativity of body-self-in-the-world. If the ingestion of these is a necessary condition for actions that are in fact constructive and meaningful in the long run for selves and the world of which we are members, they are not addictive, I believe. The danger, of course, is that these chemicals will delude people (and perhaps their psychiatrists) into thinking they are acting constructively when they are not. If so, body-selves will not be engaged in the actual tasks and challenges of the world, will not be exercised and grow.

It must be annoying, perhaps depressing, for many scientifically minded people when (and if) they discover that spectacular and important discoveries in neuroscience do not answer fundamental questions of the human condition. Under consumer and secular society pressure to expedite our desires—in effect to sanctify them in a world devoid of traditional sacral satisfactions—and incited by technology, the urge to silence fundamental questions about regeneration, self, and addiction is strong. Some very real human problems can be cured with new drugs, and the siren call to "design your own brain" is in the air. In a society set to expect quantum leaps of scientific and technological advance, the impulse to reach for the quick fix is sometimes irresistible, or nearly so.

For example: recently researchers announced the discovery of the first drug that seems to improve working memory. Again, the discovery sprang from work on neurotransmitters and their receptors in the brain, "shapely molecules that neurotransmitters fit like keys in locks."[21] "The drug BDP binds to receptors for the neurotransmitter glutamate, which triggers neuronal changes that constitute memory."

Sharon Begley asks, "Who could criticize a drug that stamps the rules for long division into your child's head after a single lesson?" But long ago William James spoke of the dangers of premature crystallization: satisfactions so great and early that one is fixated on them, and fails to achieve the greater satisfactions that might have come from long work and more risk. Sigmund Freud later noted that fixation can as easily come from too much satisfaction as too little.

The fact is, drugs like BDP raise vexing questions about self and world. On many levels of body-self functioning, humans bind together past, present, and future. The sense of self is essential to the being of self, and this sense grows through the felt ability to exercise choice, responsibility, and sometimes gritty endurance in binding together the time of one's life. Life is dramatic and suspenseful, noted Dewey: ends-in-view may or may not be realized. The self grows through its ability to engage in projects, to achieve dramatic and regenerative unities of experience—beginnings, middles, ends; even efforts to achieve these tend to be regenerative. Thus the way is opened for fresh ecstatic growth—deeper intermingling of roots in sources and branches in possibilities.

The mythic image of the World Tree suggests itself: for each of us body-selves the experiential center and axis of the world is like a tree that roots in the dark earth and intermingles its branches in the heavens and holds everything together.

What *is* ecstasy that is regenerative over the long haul of a life? We face, yet again, the contemporary dilemma of the triumphs of science and technology outdistancing the standards needed to evaluate what we are doing. Begley quotes the psychiatrist David Luchens, "If we have something that made people unshy, are they obliged to stay shy because of some ethical concern? What's the difference between 'I'm unhappy because I don't like my looks' and 'I'm unhappy because I'm shy'?" Begley rejoins, "For openers, one's core being is defined more by character traits than by the shape of one's nose."

Character traits and core being! The scientistic mentality tends to regard brain drugs as merely correcting for a metabolic or neurochemical defect. But character is affected. How do we pin down and measure character in a way that is compatible with conventional medical materialism? How could we precisely measure the ecstatic intensity of taking responsibility? The concept of body-self itself eludes medical materialism.

Drug addiction is just the most obvious way of trying to satisfy the need to *be* without acknowledging what it entails. The addict tries, childishly, to get the rewards without doing the work. In the end this destroys character and real personality. William Burroughs writes from extensive personal experience with drugs, but has little to teach us about what he calls—somewhat reductively—"the envelope of personality":

> Doolie sick was an unnerving sight. The envelope of personality was gone, dissolved by his junk-hungry cells. Viscera and cells, galvanized into a loathsome insect-like activity, seemed on the point of breaking through the surface. His face was blurred, unrecognizable, at the same time shrunken and tumescent.[22]

An arresting but partial description with no hint of a prescription.

Breaking the Spell of Addiction

The protean, ambiguous, polyvalent body: its urges range from "wholly me"—or "I-myself desire this"—typical for some persons, through intermediate possibilities, to the mocking and befuddling "me and not me" of those in the spell of addictive cravings.

If the addictive urge is felt to be just a physiological event, there are endless causes that can be taken to account for "a mere effect." Science become scientism facilitates this addictive evasion—thus the towering difficulty in helping addicts to confront their responsibility and their guilt, particularly today. When they find one cause to be phony, they immediately take refuge in countless others. And as long as gratifications are available that numb and distract from the guilt, the energy of evasion is practically limitless.

The best chance of penetrating encysting layers of self-deception lies in somehow facilitating experiences of ecstatic responsibility and character. This may tip the balance to the "I and me" side of the equivocal state, and may do so long enough to restore the cycle of the natural highs of primal needs being met by primal satisfactions that occur within the regenerative rhythms of Nature herself. Only a new, regenerative habit can displace childish ones of addictive "adaptations." How to acquire such a habit?

One hope is to experience a holy fear of the fathomless altogetherness of things, the wilderness world, which can supplant neurotic fear and desperate substitute attachments. A holy fear or awe in which we honor our hunger to belong to the more-than-human domain of living and non-living things, and endure in satisfying it. Though at times maturity dictates saying no to desires, this does not exhaust the role of character. It includes bonding sacramentally with sources in Nature, as well as the courage and humility to recognize equivocal and disorienting "me and not me" experiences. Addictions are trances, tunnels of constricted alternatives: either satisfaction *now* or anguish. Only in moments of grace that come like miracles does awareness dilate to include other possibilities. It's as if an unimagined horizon opens and the spell is broken.[23]

More on Issues of Character and Ethics
Impressed by scientism, and unwilling or unable to confront the thicket of human identity, we are tempted to dismiss ethical concerns as merely the worries of the timorous over lost status quo. This dismissal is part and parcel of what I and others take to be a pervasively addictive society.

The grounds for ethical concern become clear if we compare psychoactive drugs to, say, eyeglasses, which were vilified not too long ago as the work of the devil. The use of eyeglasses, however, was assessable within available frameworks of evaluation that are traditional and delimited. We can see through the eyes as well as see them, see their functioning, and know when this is improved. The use of glasses does have side-effects: people look different when using them, perhaps closed-off and defensive in a sinister way. But soon most observers will agree that users are no more diabolical than anybody else.

It is a different matter with drugs that directly affect neurotransmitters. Ordinary humans cannot see the nervous system and know when its long-term functioning is improved. Standards of evaluation aren't available, neither to scientists nor nonscientists.[24] Long-term side effects of neurotransmitter-manipulating drugs will be unimaginable, the effect on character real but unevaluable. Science become scientism facilitates addiction.

The scientific and also the academic-philosophical mind might listen to ordinary folks speaking out of the obscure but real depths of their lives about sources of ecstatic meaning and vitality:

> Although many patients are thrilled with their Prozac personalities, some find it disconcerting to be shielded from their sorrows. Jackie McMann of Los Angeles (not her real name) took Prozac while mourning the death of her twenty-one-year-old son. "I felt better," she says, "and my friends saw an immediate difference. But I still don't like to have my mind altered. It short-circuited my grieving." . . . Dr. Keith Ablow, a Boston psychiatrist . . . describes . . . a patient who suffered terrible abuse as a boy, doused himself with drugs and alcohol as an adult and wound up seriously depressed. His mood quickly brightened when he started taking Prozac. Yet as Ablow tells it, "he worried that he was living out his life without a core understanding and might never be able to gain it." Painful as it was, the patient went drug free and has never regretted it.[25]

Will humanity get to the point where we no longer need "a core understanding?" Would this be progress? Will we drift—or catapult—so far from awareness of primal needs that we no longer recognize them, and can only restlessly follow the endless tracks of addictive counterfeits? To be able to recognize and understand—stand under and bear—even a fracturing of the self, regardless of ongoing suffering, is to regain the self ecstatically and regeneratively. It is moral character.

The chorus of Sophocles' *Oedipus Rex* chants: " 'Tis Time, Time, desireless, hath shown thee what thou art . . ." Part of the problem of addiction is the good will of physicians and physiologists. If they can cure a person's apparently hopeless addiction or depression, they want to do it *now*. But if they succeed, the inference may be drawn that *the* cause of the malaise has been eliminated. Yet it may be only the immediate or proximate cause or condition. Thus the inference masks out the larger picture: a whole way of being-in-the-world.

The "mental set" lying behind the fallacious inference is typically a materialistically skewed dualism that conceives that "matter" causes "mind," but not the reverse. Mechanical, linear causation is simply assumed. It does not conceive that the meaning of a total situation for an agent involves continuous brain activity that must be understood as much in terms of this ecstatic meaning as in terms of the brain's own physiochemistry. To repeat, I think that "mind" and "matter" are but two aspects of one energic reality. A defective brain state may be merely the obvious aspect of a reality that includes a more concealed aspect, a "mental condition" in which the world appears typically warped or stifled to an individual or culture.

Is it fantasy to suppose that many people today are viscerally appalled at the destruction of wilderness and our natural habitat—Mother Nature cycling awesomely? I suspect they feel Earth hurting. The instinctive feeling of themselves as part of Earth prompts a corollary in their brain's functioning. That part of Earth that is their brain malfunctions.[26]

The lights in the greenhouses of the environmental college are kept on all night. Doubtless this is necessary for certain experiments that disclose things about the life of plants not disclosable otherwise. But no matter how much startling data is produced in this manner, it cannot be sufficient for our needs. It cannot help us confront possibilities and make decisions about treating the plant life of the planet to best enhance our communal life with our fellow living beings. In fact the data may hinder us, may occlude that memory of awakening at dawn in a primal forest dripping with dew and glistening in the sun—that incredibly fresh and refreshing place and time.

Ecstasy and Empiricism

Loosely construed, the meaning of *empirical* is simply "known through sense experience." But since the mathematical-scientific revolution of the seventeenth century, the stress has been on sense experience channeled through measuring devices. It must be mathematically precise sense experience. We construe what is to be studied as a variable that varies precisely relative to others.

Contemporary examples employing empirical concepts are 44.55 foot-pounds of force, 3.5612 light years, or 36.1 calories of heat. Many think a scientific account of the world is what we must take it to be given science-technology's instruments for observing, manipulating, and measuring it experimentally. Science *is* mathematical-experimental science, and it is *the* reliable account of the world.

However, the concept of "the empirical" is not an empirical matter. We cannot weigh it or measure it or experiment with it as we can with the calories of heat that a radioactive object generates, for example. No degree of mathematical exactitude delimits the scope of "empirical."

Ecstasy is empirical in the ancient experiential sense. It cannot be precisely measured, hence tends to be occluded. Only an aspect of certain subspecies of ecstasy can be precisely measured. Thus, for example, the famous Masters and Johnson study of decades ago measured conductivity, temperature, and blood-engorgement of the sexual organs during orgasm.

But this approach completely overlooks the many dimensions and possibilities of ecstatic experience. We must make a tripartite distinction between ecstasy as sensations, as emotions, and as moods. The Masters and Johnson type approach focuses only on the physiological aspect of a certain kind of sensation. Sensations, emotions, and moods each exhibit their own type of ek-stasis—standing out and in. In the moment of orgasm one may be fused with, identical with, one's sensational body; all distance between body-self and any object, even itself, may be overcome—for the moment (and it may not be overcome, as in dismaying premature ejaculation). It is ecstatic "transcendence" of its own special kind.

Emotions, on the other hand, involve various perturbations of the body certainly, but emotions are what they are because of what they are *of* or *about*, and they needn't be of or about the body. There is no loving without something loved, no fearing without something feared, no hating without something hated.

Moods, finally, must involve the body too. Depression is typically marked by a typical bodily posture, elation by another. But we needn't be able to say what we are moody about. Most moods are so totally pervasive that we can only say something like, "It's a bright and happy world," or "It's a rotten world—why was I ever born?"

Our ecstatic complexity is awesome. It is possible to have intensely pleasurable sensations and simultaneously—or nearly so—very negative emotions about these very sensations. Or possible to have both positive sensations and emotions and a vague but profound undercurrent of moody negativity and anxiety about the whole state of affairs. Or still other combinations: negative emotions and sensations and yet a positive mood of steady faith that all will yet be well—the mood of the person of character.

It is simply impossible to measure anyone's total, multidimensional ecstatic state at any particular moment. Yet ecstasy is as concretely and experientially real, or more real, than anything that can be imagined.

Failure to acknowledge this is another feature of an addictive culture, in which the "sure thing," sensation-ecstasy *now*, tries to substitute for the full dimensions of ecstatic capacities open to persons of character.

Some Empirical Illumination

Does involvement in regenerative wilderness Nature tend to ameliorate addiction? It is clear why "hard" empirical researchers are wary of rushing into such an investigation, and why in fact there has been a paucity of attempts by modern empiricists of any stripe even to imagine such an hypothesis. Hard scientists, modeling themselves on mechanistic seventeenth-century physics, would probably say our thesis pertains to a greatly complex situation that holds too many uncontrollable variables for experimentation.

But the catch in that response is the same as in all "problem approaches" to learning: It suggests that in principle there must be some way of conceiving what would solve the problem if that solution were to be found (with conceiving limited to hard, natural-scientific terms). But I do not believe we can precisely specify ecstasy and exactly how much or what kind would satisfy ecstasy deficit.

For example, what is to count as a behaviorally significant degree and duration of ecstatic involvement in Nature at any stage of our lives? One would have to be a true believer in scientism and atomism to think there must be some way to make this mathematically precise and testable.

This is not to say that all that passes as empirical investigation today is totally worthless for establishing a connection between ecstatic wholeness and wilderness Nature. Such investigations appear among the many research models and therapeutic approaches which are empirical in some sense, their proliferation and variety suggesting that addiction is not clearly understood. I will sketch some of these, not limiting myself to what purports to be hard—mathematical-empirical—science.[27] Their range speaks to the extremely broad canvas required to rough in vital connections between addiction, ecstasy deprivation, and the loss of wilderness environments in which our species took shape. I suggest that trying to achieve mathematical exactitude at every point is an unreasonable ideal—especially with regard to any particular addict. Aristotle saw certain things with great clarity thousands of years ago: (1) The educated person expects no more exactitude than the nature of the subject matter itself allows (2) There is no science of the individual as individual.[28]

One example of these investigations is that recent findings by alternative physicians tend to confirm the efficacy of natural (ancient and indigenous) modes of childbirth in which the mother places herself in a sitting position so that she can be the chief agent in normal birthing and is clearly confirmed in this. With the mother heavily drugged and supine, both mother and child are disabled. (Infants usually contribute their share by clenching and encapsulating themselves.) These physicians think that natural birthing satisfies basic needs of mother and child. Might such children be expected to be less prone to addiction than ones who weren't so birthed? How could one possibly control for other factors? Required would be a longitudinal study of probably unmanageable scope. Doubtless this is why such a study has not been made—at least to my knowledge.

In another example, psychologists and anthropologists (such as Louise Carus Mahdi and Victor Turner) document that strong selves need effective initiation ceremonies at every stage of growth. Without these, people are caught in the ever-unfinished business of trying to satisfy ill-identified needs. I do not believe, however, that these researchers have operationalized a hypothesis that predicts mathematically precise and observable correlations between lack of initiation and addiction. They observe whole cultures and note overall trends, general living conditions and levels of vitality—important work and empirical in a way. Something jibes with our experience as a whole when Mahdi speaks of prisons (usually with large addict populations) as "houses of failed initiation."[29]

Again, certain cognitive scientists attempt to pin down hunter-gatherer sedimentations in our lives, and their efforts bolster belief in the power of our intuitive, prescientific intelligence, and in our innate ability to become effective agents in the world. For example, accepting the plausible assumption that observed frequencies played a key role in primal life (for instance, the frequency of finding game at a certain place and time), we might expect that if a reasoning problem, for example, could be expressed in terms of frequencies, it might be more readily solved by us today than if it were not. Initial studies bear this out,[30] which suggests that life today more in accord with the rigors and rhythms of Nature would tend to satisfy primal needs and render us more competent and integrated, more self-respectful and less apt to be addicted. A strong suggestion, but only that.

Therapeutic Approaches and Natural Cycles

Some of the weightiest evidence for the therapeutic power of wilderness Nature that can pass in some quarters as empirical derives from studies of Outward Bound programs in eighteen countries. These launch people of a wide variety of ages into strenuous and often frightening encounters in the wilderness, and report that they return with increased awareness and respect for self, others, and Nature. Addiction typically involves massive loss of respect for self.[31] If Outward Bound activities put us in better touch with our primal needs and capacities, and if addiction results from flailing attempts to satisfy poorly identified needs, then such programs should alleviate addictions. There is considerable evidence for this, not strict enough for the strictest empiricists, but meaningful for the "softer" ones.[32]

Outward Bound programs involve coordinating daily activities with the regenerative cycles of Nature: rising with the sun and settling in for the night at sunset, great exertions followed by periods of profound rest, driving hungers or thirsts and great satisfactions of allaying them in due course, etc. Would benefits be greater if this correlation were more explicit and developed, and included monthly or seasonal natural cycles, say, or explicit recognition of the generational cycles of leaders and the young people who will some day take their place? I don't know if much is made of such questions, but much might be.

Consider now a program for treating addictions that is highly compatible at certain points with the conceptual framework being developed in this book. It is detailed in Rozanne W. Faulkner's *Therapeutic Recreation Protocol for Treatment of Substance Addictions*. Intriguingly, Faulkner views addiction as "leisure malfunction," as an "abuse of leisure." This characterization implies that optimal stress/restoration rhythms are essential for regeneration, health, maturity. There is a need for restoration, but addicts don't know how to satisfy it. They typically *define* leisure as a fix. Faulkner aims to restructure their lives by restructuring the time of their lives—re-rhythming them, I would call it.

She notes trenchantly that members of an alcoholic family typically show an "inability to have fun."[33] Whether addicts are referred to her by the courts or sign up on their own, they are all required to attend scheduled classes in which they learn recreational (re-creational) skills—drama, various arts, games, swimming, gardening, fishing, adventure and risk taking, and more. The goal is to teach them to be "leisure independent," that is, to be able to enjoy leisure time as intelligent agents, and to be thereby restored.

Faulkner regards "adolescent scheduling" as crucial. She addresses in particular the problems of "latch-key children," those who come home from school with no adult there to greet them and to ritualistically confirm them in their transition in the cycle of their day. They easily fall into the habit of quick, stop-gap gratifications. Her case histories are revealing, and some attest to her success in difficult cases. For example, "Tina Bopper" was a fifteen-year-old runaway who had become a prostitute and drug addict. Tina had to learn how to play and "to grow from being a woman back to being a child" so that she could retrace the stages necessary to build a responsible individual self.[34] Some solid success was achieved. If we are to explain how this is possible we are challenged to imagine how human beings live in time ecstatically. (I discuss human reality and time in this book's chapter 9.)

Significantly, Faulkner recognizes, at least tacitly, daily cycles and the opportune times of the day to schedule exercise and recreation. Also weekly cycles: on Wednesdays and Saturdays there are special events, which seem to be occasions for holiday restorations. Whether monthly, seasonal, and generational cycles are also recognized is not clear.[35]

Alcoholics Anonymous

For Alcoholics Anonymous, engagement in Nature is not central to the program. Igor Sikorsky in *AA's Godparents* has traced the history of this group and its redoubtable twelve-step method. What stands out immediately in his account is bonding in human community and fellow-feeling (a primal need that hunter-gatherers must have usually satisfied). Relationships are formed that aim to be as reliably supportive as those in a good family or hunting band. Some "couples" call each other every day or every other like clockwork. Members share a corporate identity, "recovering alcoholic," no matter how long they have been sober, for they know there is always the possibility of relapsing.

Above all, they recognize that by their own efforts they cannot succeed, that without confessing inadequacy and accepting gifts from a "higher power" they are impotent. This adds a clearly sacramental element in the de facto ritualized life with its mantra or chant, "one day at a time."

The mantra acknowledges the fateful human tendency to overextend the ecstatic life of imagination and ambition and lose touch with the here and now—the concrete situation, one's body-self always embedded in it, as well as the tendency to deceive oneself about what's closest at hand. All of AA's efforts presuppose honesty: "a searching and fearless moral inventory of ourselves" (step #4), the steadfast refusal to get lost in airy abstractions, grandiose projects, and countless modes of self-deception.[36]

How successful is AA? Herbert Fingarette notes that the group has long been reluctant to gather or publish statistics.[37] He also observes that AA is highly self-selective—comprising only about 5 percent of all alcoholics—and that the recovery program is simply not attractive to a majority of problem drinkers. Most never join, and many drop out once they do. Some reject the religious overtones ("the higher power"), others the disease concept embraced by AA: the belief that the addiction is a grave "allergy," that drinking cannot be controlled and total abstinence is the only answer.

As Fingarette argues, AA works only for those who can embrace it as a vocation, one that weaves them into the world in a way that establishes their significance and confirms their character. I believe he is right, that finding excited significance in one's self as moral agent is a primal need, and that other needs may bend in the wake of it, even the need to take a drink. We can't say that we simply respond to "objective situations," even our own "objective needs," for choices of ways of life heavily influence how situations and needs are interpreted, which influences what they are.[38]

I have already called attention to the ambiguity of "higher power" and of the dangers of quitting drink but keeping an addictive style of life, "heroically transcending" primal bodily needs. Charlotte Davis Kasl asks us to rethink the entire twelve-step program.[39] It was evolved within a culture based on patriarchy and exhibits, she believes, limits of that whole heroic approach that are concealed from the approach itself.

She advances a sixteen-step program that recasts recovery from a female point of view. The steps center around her insight that women needn't repeatedly declare their powerlessness: they are already disempowered by the culture. Step by step (for example, step ten: "We learn to trust our reality and daily affirm that we see what we see, we know what we know, and feel what we feel. . . ."), she moves to step sixteen: "We grow in awareness that we are sacred beings, interrelated with all living things and, when ready, take an active part in helping the planet become a better place to live."

As we will see in chapter 11, when a supreme being is thought to be female, everything looks different. The meaning of creative stress and relaxation and ecstatic growth shifts, as does the meaning of individual self: In participating more effectively and spontaneously in the Whole, individuality is not denied but rather enhanced.

Reconfiguring the Imaginal and Conceptual Ground

Paul Shepard's unabashed appreciation of wilderness experience and his penetrating assessment of the costs of its loss are foundational, I think, to any comprehensive view of addiction, and of empirical studies and therapies. He adroitly employs a range of psychoanalytic and general psychological findings. He considers the effects of crowding in cities, even though realizing that he runs into the question of "exactly how much is deleterious?"[40] This does not deter him. He reminds us how the geometric, isotropic grid of seventeenth-century mechanistic physics is monstrously concretized in contemporary cities. He cites radically reduced flora and fauna, the absence of wilderness and its natural cycles of sleep and waking, and frustration of the primal need to explore and orient ourselves in the ancient bodily ways: by the landscape and the horizon, the four cardinal directions, the seasonal movements of animals and birds, and the majestic movements of heavenly bodies. In isolating ourselves from wilderness conditions, Shepard maintains, we isolate ourselves from ourselves:

> All children experience the world as a training ground for the encounter with otherness. That ground is not the arena of human faces but whole animals. Nonhuman life is the real system that the child spontaneously seeks and internalizes, matching its salient features with his own inner diversity. . . . The city contains a minimal nonhuman fauna. Adequate otherness is seldom encountered. A self does not come together that can deal with its own strangeness, much less the aberrant fauna and its stone habitat . . . The world to the child—and adult—is grotesquely, not familiarly, Other.[41]

When swimming with the salmon I felt ecstatic kinship with strange beings. The strangeness, stretching and expanding the self, was essential to the ecstasy. The fish were fearsome, other, but compatibly alien.

Their recognition of me as a fellow living being was a stirring visitation in my body of the sacred togetherness of things. They satisfied a primal need—one exploited by Disney and subverted into the cozy self-deception of humanoid ducks, dolphins, rabbits, mice, and lions. Mickey Mouse and his cohorts cannot sustain us beyond brief periods of distraction, and if more than this is not found, the distractions will be repeated addictively.

My dog was a domesticated animal, but a real one, and he led me toward wild ones, such as the fox. If I could find myself in the vast community of beings I could find all of myself, and own up to all of myself. I would be most my individual self because most a member of the World-whole, a member recognized at the fundamental level of wilderness experience. If my life is valuable just because of itself, there is no need to exploit it in ephemeral gratifications.

The Limits of Empiricism

Given spotty success in dealing with addiction culture-wide, medical materialists press all the harder on chemical-therapeutic approaches to addiction; these are legion. Consider psychiatrist Marc Schuckit's *Drug and Alcohol Abuse: A Clinical Guide to Diagnosis and Treatment*, now in its fourth edition.[42] As one would expect, it details far-flung, varied, and intricate pharmacological procedures in the most mathematically precise physiological terms. But late in the book, at paragraph 15.1.1.9, he writes, "Because there is no "magic cure" in these areas, recovery is usually a long-term process that requires some counselling and therapeutic relationship for at least six months to a year." That exhausts the paragraph, and in 15.1.1.10 he writes,

> The patient's substance-abuse problem does not occur in a vacuum. Part of the therapeutic effort should be directed at encouraging the family, and, if appropriate, the employer to increase their level of understanding of the problem; to be available to help you whenever necessary; to make realistic plans for themselves as they relate to the patient; and, in specific instances, to function as "ancillary therapists. . . ."

Dr. Schuckit has confined his study, not even trying, apparently, to define addiction. He notes in passing on page three that

> the compulsion surrounding some forms of gambling has much of the "feel" of the obsessive behavior observed during substance abuse. However . . . expansion into these topics might jeopardize my attempts to

cover clinically related topics succinctly and thereby help the clinician in his or her day-to-day practice.

What price succinctness? In passing over addictive gambling he ignores the evolution of the species that conditioned us to expect risk and suspense. Many today crave these, to the point that gambling mangles their lives.

But his momentary dilation of vision at the end of his book is revealing. It's as if he were conceding the limits of "empirical and problem approaches" to addiction through medical technology, as if he were lamenting the absence of a rooted community with effective rites of passage within an intercommunicating world: sparkling rites of the maturation of human life and of the ecstatic formation of individual character.

Certain scientific-empirical approaches are suggestive. But each of us is an individual with ecstatic capacities that cannot be measured; each of us can be specified in only the roughest, open-ended way. The ultimate snare of addiction is believing there is a quick fix for understanding it. Or that we needn't understand it, but only try to stop it through massive police efforts.[43] The approaches outlined above detect pieces of the picture, but needed is a more expansive, imaginative, and organic vision of how these might fall together into a whole.

Notes

1. M. M. Glatt, M.D., *A Guide to Addiction and its Treatment*, 20 ff.
2. Recent research seems to contradict this. Two complex experiments by "hard" scientists trace the symptoms of emotional distress caused by marijuana withdrawal "to the same brain chemical, a peptide called corticotropin releasing factor (CRF), that has already been linked to anxiety and stress during opiate, alchohol, and cocaine withdrawal" (*Science*, 276, June 27, 1997, "Marijuana: Harder Than Thought?").
3. Reported in Howard T. Milhorn, M.D., *Chemical Dependence: Diagnosis, Treatment, and Prevention*, 3. See also M. M. Glatt, M.D., cited above.
4. 10.
5. Of course, one might get drugged and be unaware of it, and become "physically dependent" in some sense. While undergoing an operation, morphine might be administered, and when it wears off withdrawal symptoms alleviated through another dose of it. But it would be straining terms and concepts to call such a person "an addict"—unless they took it up as a style of life.
6. But they might. Is there a typical biochemical correlate of workaholism? I do not know. For a text on workaholism, see B. E. Robinson, *Work Addiction*.

7. *Basic Writings*, 33.

8. Ronald L. Hoffman, M.D., building on the work of Hans Selye ("a medical genius"), describes coffee as a "stresser" to which we can "adapt," that is, the threshold for triggering the alarm bell is raised, and it is no longer heard. But all the while the caffeine "is sapping the adrenals as well as important chemicals in the brain . . . Critical minerals like magnesium may become chronically depleted. The nervous system and the immune system suffer. A person enters the next stage, called *maladaptation*." (*Tired All the Time*, New York, 1993, 26). I see no reason to doubt these judgments about physiology. I only suggest that researchers might look for some powerful compensatory effects in the total ecstatic life of which caffeine is a part.

9. *Messengers of Paradise*, 25. Developing her general view that European civilization encourages addiction to substances (e.g., gold and silver), Dolores LaChapelle writes, "addiction continues as a major source of capital and, as we run out of natural resources this will only grow worse. For example— take the . . . popular drug 'crack.' It turns out that people who've been using marijuana for years can no longer find it from their usual sources; they can only buy 'crack' . . . In four seconds it goes straight to the brain, gives a quick hit but *immediately* leaves the person with no way to produce endorphins in the brain. Totally addictive—a wonderful 'consumer' product. It makes it necessary to buy it again immediately" (*Sacred Land, Sacred Sex*, 49).

10. *Walden*, 71.

11. 7–8.

12. Ibid., 56–57.

13. Ibid., 98–99.

14. Dewey, *Experience and Nature*, 295.

15. Ibid., 38.

16. Ibid., 65.

17. See James, *The Principles of Psychology*, chap. 21, "The Perception of Reality."

18. 113.

19. Compare Rupert Sheldrake in Renée Weber, *Dialogues with Scientists and Sages*, 77.

20. Personal communication from the psychiatrist Dr. David Olds, fall 1993.

21. Sharon Begley, "One Pill Makes You Larger, One Pill Makes You Small," *Newsweek*, February 7, 1994, 37–40.

22. *Junky*, 58.

23. On the breaking of spells, see Buber, *Between Man and Man*, 3–4.

24. In the case of obvious pathologies such as schizophrenia, standards are fairly obvious. The patient cannot concentrate on what is essential in a situation, must "try to pick something out of the whole room," as one said. Regulating norepinephrine in the brain may clearly help. I am mainly concerned with the use of neurotransmitter-manipulating drugs in the general population.

25. Geoffrey Cowley, *Newsweek*, February 7, 1994, 42.

26. On the holographic metaphor or a model of how parts of the brain may

exhibit the whole in some way, see Pribram, *Brain and Perception: Holonomy and Structure*. I suggest that that part of the environment which is the brain may exhibit the whole environment to some extent. Consult David Bohm's idea of holomovement in four interviews in Weber, *Dialogues with Scientists and Sages*.

27. As "hard" science, empirical theory makes predictions testable in empirical terms. See Hartig et al., "Perspectives on Wilderness." This reviews "hard" attempts to evaluate restorative powers of wilderness as well as new efforts to tighten their rigor. They acknowledge their hypotheses derive from the Kaplans and Kaplan and Talbot ("Psychological Benefits of Wilderness Experience") who build on James's idea of involuntary attention and fascination and implicitly on Emerson's idea of horizon. Fascination requires coherence if its not to be confusing (87). Note "distance coherence": "a sense that there is continuation of the world beyond what is immediately perceived," and how this encourages exploration. Employing the rigor of social science trying to emulate natural, Hartig et. al. predict measurable restorative consequences of wilderness over urban experience, and claim these are fulfilled in the main. (At times common sense is sacrificed to exactitude: a test used for mental restoration is accuracy in proof-reading. But who would want to do that after a wilderness trek?) A "hard" attempt to confirm restoration from addiction would have to generally follow their methods.

28. (1) *Nichomachean Ethics*, Bk. I, chap. 3, lines 24–26. (2) This theory appears in many places throughout his writings, for example, in *Physics*, *The Organon*, and *Metaphysics*. Science deals with classes of things, so each individual as individual must escape final specification by science.

29. Mahdi, *Betwixt and Between*, xi.

30. Stephen Stitch, lecture, Rutgers University, November 30, 1995. A body of presumably scientific literature begins to emerge on innate intelligence, for instance, G. Gigerenzer, "How to make cognitive illusions disappear: Beyond heuristics and biases," *European Review of Social Psychology*, vol. 2, 1991; also that author's "The bounded rationality of probabilistic mental models," in Manktelow and Over, eds., *Rationality: Psychological and Philosophical Perspectives*, London, 1993. Also L. Cosmides and J. Tooby, "Are humans good intuitive statisticians after all?" . . . to appear in *Cognition*. Finally, Steven Mithen, *The Prehistory of the Mind*.

31. Mark Zelinski, *Outward Bound*. Also B. L. Driver and associates, as well as R. and S. Kaplan on Outdoor Challenge Program research, *The Experience of Nature*, indexed.

32. For example, B. P. Kennedy and M. Minami in "The Beech Hill/Outward Bound Adolescent Chemical Dependency Treatment Program" (*Journal of Substance Abuse Treatment*, 10 (4), July–August, 1993, 395–406) report on a study of adolescents with substance abuse disorders (aged 14–20) at an inpatient treatment facility, which program involved a twenty-two-day wilderness experience. Phone interviews with each adolescent and a parent followed this up at three-month intervals. MMPI and Personal Experience Inventory data were collected along with detailed psychosocial assessments

to decide which factors predicted successful outcomes. At one year post-treatment, 47 percent reported complete abstinence from alcohol and other drugs. Particularly with respect to family relationships, interpersonal life was markedly improved. Studies were made at various levels for severity of drug use and psychopathology and for attendance at self-help groups such as AA. Subjects with severe psychopathology and drug use scores who were not attending AA were 4.5 times more likely to relapse than subjects with low scores who attended AA. ("Hard" empiricists will object to "lack of controls": "How much is due to AA? What value is 47 percent abstinence when we don't have control groups who try other means? Or no means? Some may just mature-out of addiction.") Other studies corroborate this general finding, though not with such apparent exactitude. See J. D. Mc-Peake, et. al., "Innovative adolescent chemical dependency treatment and its outcome: A model based on Outward Bound Programming" (*Journal of Adolescent Chemical Dependency*, 2 (1), 1991, 29–57). They report that most of the adolescents contacted six months, one year, and two years after treatment had significantly decreased their substance use, were happier with themselves and others, and led lives that were more productive. Also see H. W. Marsh et al., (*Personality and Social Psychology Bulletin*, 12 (4), 475–492, December, 1986). Their report in "Multidimensional self-concepts: A long-term follow up of the effect of participation in Outward Bound Programs" suggests that self-concept can be changed and that these effects can be maintained. Also see D. R. Busby. In "A Combination that Worked for Us" (*Federal Probation*, 48 (1), March 1984, 53–57) he used the North Carolina version of Outward Bound, and non-statistical evidence of its effectiveness is described. Also see doctoral dissertations on Outward Bound in general: J. Tangen-Foster, Dissertation Abstracts International, 53 (6-B) [December 1992]: 3212–3213; R. D. Boudett, DAI, 50 (11-B) [May 1990]: 5306–5307; D. J. Jordan, DAI, 50 (5-B) [November 1989]: 1885–1886.

33. 6.
34. 219.
35. Also see the Ditzlers, J. M. and J. R., "Treatment at Farm Place," *British Journal of Addiction* 84 5 (May, 1989): 493–497. Exercise is prominent, as are some of the rhythms of farm life. There is a focus on eating disorder addictions, and on others.
36. The steps, along with extensive commentary, can be found in T. T. Gorski, *Understanding the Twelve Steps*.
37. *Heavy Drinking*, 88 ff.
38. 104.
39. *Women, Sex, and Addictions*.
40. *Nature and Madness*, 94.
41. 98.
42. 306.
43. Is the United States, the richest nation ever to appear on Earth, rich enough to succeed in this? Doubtful!

Chapter Seven

Possession, Addiction, Fragmentation:

Is a Healing Community Possible?

The nonaddictive body-self viscerally acknowledges and makes its own its primal needs and urges as an ecstatic being-in-the-world. It possesses itself. When I ran with Buppie in the fields and swam with the salmon I was coherent ecstatically. Their awareness of me felt by me strengthened my alertness and individuation within the Whole. The experience was valuable and satisfying in a way that unutterably vouched for itself.

But what exactly is this ecstatic coherence, this oneness? Any individual, no matter how fractured in addictions, is called one individual. Indeed, each is one being, but only as understood for our everyday—even scientific—purposes of classification and control. When we try to grasp human identity as we actually experience it through time, we find that the assumed oneness often papers over and conceals addictive cravings experienced mockingly as "both me and not me." What remains is numbed repetition and baneful trance.

Thinking of the person as an individual, as one, the thinker easily slides into thinking of him or her as like the number one: a unit homogeneous through and through and simply identical to itself. As $1 = 1$, so "I" = "I", Ego = Ego! As if identity of self were merely an abstract one like a number's!

The danger of oversimplifying human identity by naively contrasting self and other ("I'm one, you're another, that makes two"), inheres in all mathematical approaches, particularly in all that are also dualistic. The mind is supposed to be one mind, and essentially reflective of its "mental contents," as if it were a mirror-lined container. We might define the self as that which says "mine" of itself. It is then tempting to infer that since the mind does this, *it* therefore *is* the self, or the core of it.

But once "the mind" is exposed as a verbalism parading as concrete reality, the impression evaporates that anything important about self has been learned. Defining self in this way is no better than saying I am who I am, identical to myself, because there's a little man inside my body—me—who is identical with himself. But what makes the little man identical with himself? There's a still smaller man inside him who is identical with himself. And smaller and smaller men inside men—homunculi—are supposed, ad infinitum. This gets us nowhere.

I think that the self is in fact a body, the inside of which directly experiences itself resonating to the rest of the world, and which as a being that moves itself contrasts itself frequently to everything else in the environment. Without this frequent contrasting there could be no personal mineness, no sense of myself. The inner body's primal sense of mineness is what ought never to be violated or infested with otherness that cannot be made one's own. When we possess continuously a visceral sense of "my own body-self," mocking "me and not me" cravings are impossible, crazy-making cravings cannot happen.

Even when we are bogged in addiction and guilt, however, the call to the self of maturity and freedom is seldom completely silenced. We may at least be free to remonstrate with ourselves, to persist in a state of dissatisfaction that might yet prompt a leap of abandonment in some greater matrix of energies. A wild bird may flash into a life, or a dog, or a fish, or a dawning day may awaken one to startling possibilities . . .

Redrawing the Conceptual Map

The initial strokes of a new conceptual map have already been laid out in chapter 4. The body-self is primally individual because it distinguishes its movements from the rest of the world's. But this is to be understood within James's idea of pure experience, at a level more fundamental than that of the simple oppositional distinction between self and other (or mind and matter).

We must now better grasp the envelope of the body not just as skin that moves relative to other things in the environment, perhaps touching them, but as that which encloses and locates its own innards—dark inner cavities, organs, fluids, subvocal speech—and provides much of the basis for the sense of "my inner self." We must grasp this envelope well enough to understand the loss of self in demonic possession, and its impairment in demonic infestations, addictions—impairment captured in these overheard words, "You couldn't have called her that, after all she has done for you. It must be the booze or the drugs talking."

I argue that addiction is a function of vulnerability of body-self, of our inability to trust the world to respect and nourish the integrity of our "inner self." I mean our inner self as an individual body immediately experiencing itself from within itself, experiencing basic periodic needs: for ingestion, digestion, excretion, exploratory initiatives; for sexual activity; for centeredness and poise that can found initiative; for this innerness to be recognized by all as that which makes each self fully human. Above all there is the need to feel one's life to be significant, to be fully one's own, the need to which other needs may be sacrificed.

Rituals

Addiction is the failure to stand trustingly open to circular power returning periodically and regeneratively into itself through ourselves—body both suffused by the environment yet able normally to contrast itself to it.

How does the self achieve integrity as an integral member of the ongoing regenerative world? Only through rituals, I think, such as initiatory ones at crucial stages of development, that weave a moment of our life into the abiding and timeless ground of our being in the world. Ritual interweaves the two temporal dimensions of our lives: the linear on the one hand, and the eternal present, or cyclical, on the other. Like all events, any particular performance or performing of a ritual happens at an unrepeatable point in time. But what is performed is the very same event reactivated: that which established and repeatedly reestablishes identity in individual and group (An archetypal example, "At thirteen all Jews have been, and will always be, inducted into adulthood.").

Such rituals operate at the level of origin myth. By "myth" I mean just the opposite of "false" or "fantastic." For only by the means of myth can we be true persons, solid and coherent (see chapters 9 and 10). Myth is the unquestioned sense of the unconditional power and value of the Sources that have generated and do regenerate us and cannot be effaced by time, that have names that should be capitalized. This sense of our Sources is sustained in ecstatic ground moods that stabilize us.

Only rituals based in origin myth secure the intrinsic value and genuine self-regard of the person. They affirm us as most individuated when most in tune with the Whole. This is an obvious corollary of the general principle of value. Ritual is the deliberate intensification and clarification of ourselves.

It is impossible to mark sharply where ritual begins in everyday life. Is brushing one's teeth a ritual? If it were merely a repeated behavior that could be understood in completely utilitarian terms—brushing deposits off the teeth—then perhaps not. But this is probably never altogether the case. For nearly always some statement is being made in repeated behaviors, even if only to oneself—in the tooth-brushing case it might be, "I am a conscientious and cleanly person"; and the behavior tends to reinforce these character traits through time.

Every performance of origin myth is obviously ritualistic. When one reaffirms one's place as a member of the World-whole, the deepest need is satisfied—to *be* vitally.

Given the use of drugs in our scientistic and secular society, it is easy to think that all use of drugs is addictive and degenerative. When, however, we turn to an indigenous group such as the Yanomamma of Venezuela, we may think differently, for everything these people do is done within the context of origin myth, of the "spirits" that generate and maintain their World. Anthropologist Kenneth Good writes,

> The purpose of drug taking . . . was to put them in contact with the spirit world. The shamans took the drug for that specific reason and for that reason only. Other men could also take it, and they did, but only by way of participating in the shamanistic chant. Drugs were not taken just to get high. Without a ritual context and purpose, drug taking would seem a foolish activity.[1]

Moreover, these people recognize excessive or deviant use of drugs:

> Taromi is intoxicated with yakoana drug, as he is every day . . . Ebrewe looks at him with a contemptuous pout and delivers the following judgment: "He inhales only the yakoana drug, which ages you before your time when taken to excess. Look at his buttocks, already furrowed and slack."[2]

Demonic Possession and Addiction
Only when we grasp the absolute importance for the integrity of body-self of regenerative rituals can we grasp how their breakdown may entail demonic possession—and its seemingly milder but more insidious form, demonic infestation, addictions. For body-self is not a homogeneous unit over against the otherness of the world, but an ecstatic body with a crucial kinesthetic sense of its own insides as they resonate with the wide world. If this wider reality is untrustworthy, is disrespectful of the inner-person-inner-body as the self tries to maintain itself, it can

become a part of the self that is incompatible, that is also "not myself."
It lies outside my approval and agency, uncontrollable. The take-over
can be abrupt and total—demonic possession—or creeping and insidious, as when one speaks of an addiction, "That can't be you talking, it
must be the booze." Even our own bodily urges can be experienced as
thing-y, incompatible, if we cannot appropriate them whole-heartedly
as our own.

Any uninvited incursion into the body, threatened or realized, is not
just a violation of the body, but a violation of the self. Even one's own
body's urges may be experienced as incursions. To repeatedly serve
them is to become servile, polluted. Addictions can be understood as
degenerate rituals that not only fail to prevent pollution or infestation,
but constitute forms of them.

Here Mary Douglas's monumental *Purity and Danger* should be con-
sulted. She analyzes ancient Hebrew purification practices integral to
their origin myth. These embody the belief that clean things live and
move properly in their environments, while dirty things do not. Eels,
for instance, have no fins and move improperly in the water and hence
are unclean according to Hebrew law.[3] It is particularly the eating of
these creatures that is polluting, for a disordered thing that enters the
body tends to disorder the self. What is disordered must be kept outside
the body.

Likewise, what is inestimably valuable if kept inside—saliva or blood,
for example—will be polluting if ejected improperly, particularly if re-
contacted or reincorporated. Pollution involves mixing what ought to
be kept separate, especially mixing materials inside the body with those
that are outside. Such mixing threatens our survival as the beings we
are.

Body-self's course is ecstatic, serpentine, vulnerable, and so can be
thrown off-balance. In some situations engulfment in intimate otherness is thrillingly joyful, in others a terrifying pollution or possession.
For persons of sound identity, engulfment in compatible and respectful
others in appropriate circumstances is consummate intimacy and ecstasy. Engulfment in disrespectful others is hell. Even in joyous erotic
relationships, when lovers probe each others' mouths with their
tongues, it is not clear that they would happily accept each other's saliva
if it passed between them in the form of drool. It would be too public,
too experienceable by others—even by themselves, separated as they
are in space and capable of objectifying themselves. When persons ex-

perience themselves as objectified and vulnerable, the exchange of bodily fluids threatens their identities, and frightens and abases them.[4]

Thus the dense wrapping of rituals in which all societies enclose sexual activity to prevent violation of self (all societies except perhaps segments of our own today). Thus the very personal and neurotic rituals that some devise to somehow deny the reality of violations in early life. Or thus just the garden variety of addictions, such as smoking in secular settings, that force a continuity of self in the face of threatened incursions and disruptions.

In the face of threat, addiction is concerted action to constrict the scope and depth of alternating ecstatic fusion and differentiation, the pulse of life. At the fully demonic stage, the alternating rhythm breaks down completely: the possessed person is clogged with incompatible otherness. In the case of addiction, body-self doesn't lose control immediately and completely, but loses it only of that element of self that cannot be made its own. An addictive craving is a fragment of self that repeats itself mechanically. It simulates integrality and continuity and tries to eliminate the possibility of demonic possession.

More on Demonic Possession

There are innumerable well-documented accounts of behavior so altered that observers believe another being has taken control of the person's body. From ancient to contemporary times, reports emerge of frail persons suddenly endowed with superhuman strength. One rips up saplings by the roots while a stench emanates from her.[5] There are accounts of once-devout persons shrieking obscenities in an unrecognizable, low, rasping voice; and of once-devoted children estranged violently from their parents. In such cases the ultimate pollution of self occurs, and the person seems to be invaded by what we call a demon.

Letting in too much alien and disrespectful otherness is the worst way in which ecstatic openness goes wrong. Recall James noting that the other's condemnation works directly in my body to produce those changes that both express and are my shame. We are mimetically engulfed in others' bodies and these bodies' attitudes. If this goes far enough, we lose the clear awareness that it is their presence as opposed to our own; demonic possession may result. And it needn't be merely human others that possess us.

I am supposing that we are body-selves incorporating other persons' and things' presence and regard. Our experiencing is not perfectly sealed off from others'. Mature selves, however, seldom wholly lose the distinctive immediacy and presence of their own individual bodies—the cubic mass we feel all the while, and its continuities of organic excitement.

Whenever threatened, the strong self can contrast its experiencing and experienced complexity, its articulable I-and-me, its I-myself, to the other being or person. Able to say "I" or "I-myself," the strong self expresses (with some degree of adequacy) its immediate experiencing-experienced reality that can exclude incompatible others. And, as I have tried to show, all this is done by a body, body as self.

Now, what happens when persons can no longer protect their privacy and make the simplest decisions about what is to be accepted or rejected? Suppose a young girl is raped by her father. His presence overwhelms her habitual defenses against incompatible and disrespectful otherness. She identifies with him unwittingly (if she could speak: "He's a loved part of me, how could this violation and destruction within myself and by myself be happening?") She can only try to flee, but can only flee from herself. All respectful distance between the one addressing and the one addressed breaks down; the other, commingling in her torn body-self, usurps her place as agent. The articulations and distinctions of the resonating body-self-in-the-world collapse. She speaks in a low, rasping, unrecognizable voice and exhibits startling strength. She is beside herself in a grotesquely literal sense: the other's view of her becomes an alien, moving, and viewing presence in the girl's body. With good reason we say she is possessed by a demon.

In such cases, the integrity of the bodily envelope has not been respected. One's ability to collect oneself and keep one's own counsel has been overwhelmed. In other words, the dynamism of experiencing-experienced self over against a world experienced has collapsed. The menacing other's experiencing is not just something experien*ced*—a distinction a strong self can make when necessary—but floods one's own experien*cing*. One is taken over by the other—some hideous, floating being that does not belong in this body.

Addictions Confronted
Origin rituals and myths authorize basic needs and appetites. It is right that we have them, right that we be. Substitute gratifications addictively repeated violate our integrity; they amount to incompatible otherness. We fail ourselves at the heart of ourselves, and must feel guilt.

Energies of evasion and denial of guilt are practically limitless because demonically fueled. Addictive life hangs on fiercely, since for it, trapped in its wheel, the only alternative it feels is withdrawal sufferings and death. In the Biblical account, a demon asked its name replies, "My name is legion." This is atomization of body-self—will, understanding, appetite fly apart. The demon would banish Jesus from its sight, for the possibility of a coherent form of life means its death. It begs Jesus to release it into a drove of hogs. Jesus obliges, but the swine stampede into the sea and are drowned.[6]

Angelic Possession

As engulfment in disrespectful and powerful others is demonic possession—a particularly grotesque version of its sibling, addiction—so engulfment in respectful and powerful others is angelic possession.

Saint Paul writes ecstatically, "I am crucified with Christ: Nevertheless, I live; yet not I, but Christ liveth in me."[7] But there is no doubt that this is the same person who was Saul before his conversion and who is now Paul—the same who was struck down on the road to Damascus by a blinding light. In this case, the discontinuity radically strengthens the self: it is angelic, anything but demonic. And since angelic possession is inherently expansive, it does not resist acknowledgment. The demonic variety and its addictive sibling are self-enclosing.

Margaret Prescott Montague revives a paganlike experience of angelic involvement in Nature as a whole:

> Entirely unexpectedly (for I had never dreamed of such a thing) my eyes were opened and for the first time in my life I caught a glimpse of the ecstatic beauty of reality . . . its unspeakable joy, beauty, and importance. . . . I saw no new thing but I saw all the usual things in a miraculous new light—in what I believe is their true light . . . I saw . . . how wildly beautiful and joyous, beyond any words of mind to describe, is the whole of life. Every human being moving across that porch, every sparrow that flew, every branch tossing in the wind was caught in and was part of the whole mad ecstasy of loveliness, of joy, which was always there. . . . My heart melted out of me in a rapture of love and delight. . . . Once out of all the grey days of my life I have looked into the heart of reality; I have witnessed the truth.[8]

Magnificent. But a primal person would be equipped with rituals that would reenact periodically one's ecstatic groundedness in the generative and regenerative world. It would not be a once in a lifetime experience.

Selves of strong character employ regenerative rituals as magnificent obsessions which invoke from time immemorial authoritative others. These are mythic beings who respectfully recognize one's individual being as a member of the Whole and thereby constitute it as such. Black Elk has invoked for him in his vision the Six Grandfathers.[9]

Running or walking with my dog—that for me was angelic possession.

Regenerative and Degenerative Rituals

The next three chapters are explorations of regenerative rituals possible for us today: smoking (ritualized, not merely addictive), walking and sensing with one's whole body-self, and art making. These are intrinsically valuable activities that can satisfy the primal wild hunger to belong in will-of-the-place and to *be* coherently. They conform to the general statement of value: the greatest individuation within the greatest coherence or unity. They radiate preciousness.

The composer Anton Bruckner, for example, suffered repetition-compulsions. But what he longed for is disclosed in his music. The entranced repetitions in the climaxes of his symphonies, and the ever-repeated cycles of stress and relaxation throughout, evoke the regenerative repetitions of Nature herself. The sterile repetitions of Bruckner's addiction—his numbering mania—aimed at something far beyond itself: the regenerative climaxes of Nature. At least in his music he discovers this. Sonic climaxes give way to silences: awe in response to the cycling world feeding back into itself, reclaiming itself.

If integrated some way ritualistically into daily living, awe evoked in the music might undermine addiction by satisfying the needs to belong to the World-whole that drive the music and ourselves. Addictions are powered by a hysterical hunger for ecstatic enlargement, for relief from nattering fears and frustrations. The hunger easily trips itself up, probably more dangerously today, when rituals have been eroded for so many people.

In periods of lucidity and industry between heavy bouts of drug addiction, William Burroughs recounts in wretching detail the course his hunger took. At first glance his story is utterly different from that of Bruckner, for there is the composer's numbering mania, but also his nearly continual creativity and devout Catholicism.

But we detect some common ground. Burroughs shows how the addict guiltily conceals from himself the hunger for heroin behind endless rationalizations. He makes rules for ingestion ("only every other day") but these allow numerous exceptions. He says, "I can quit any time," but in the agonies of actual withdrawal the will is powerless. Why the guilt, unless the addict dimly realizes he is cheating himself of real freedom, and denying the world his ecstatic involvement with it as a responsible being.

The hunger for ecstatic enlargement through drugs entangles itself. Burroughs: "As a habit takes hold, other interests lose importance to the user. Life telescopes down to junk, one fix and looking forward to the next, 'stashes' and 'scripts,' 'spikes' and 'droppers'."[10]

One must fix what is fractured. In periods of lucidity he glimpses deeper motivations nearly hidden behind his rage for the next fix to deliver from "junk sickness." Note: "a mild degree of junk sickness always brought me the magic of childhood. 'It never fails,' I thought. 'Just like the shot. I wonder if all junkies score for this wonderful stuff'."

But after making the injection, "The junk spread through my body, an injection of death. The dream was gone. I looked down at the blood that ran from elbow to wrist. I felt a sudden pity for the violated veins and tissue."[11]

What is this magic of childhood that he finds fleetingly? Burroughs is short on description.

The last lines of *Junky*:

> I decided to go down to Colombia and score for yage . . . I am ready to move on south and look for the uncut kick that opens out instead of narrowing down like junk. . . . Kick is seeing things from a special angle. Kick is momentary freedom from the claims of the aging, cautious, nagging, frightened flesh. Maybe I will find in yage what I was looking for in junk and weed and coke. Yage may be the final fix.

But his whole framework has shrunk and he within it. He remains in a drug user's world—only a better *fix* could "open out instead of narrowing down." He is trapped in infantilism.

Of course, Burroughs is writing about all this, which suggests there must have sometimes been the ecstasy of writing. Perhaps not unlike Bruckner's in the sonic climaxes that transcend the composer's numbering mania? But how Burroughs incorporated this in his life I do not know. Life is many days, day after day, moment by moment.

Notes
1. *Into the Heart*, 74.
2. Jacques Lizot, *Tales of the Yanomami*, 30.

3. 55–56.
4. W. I. Thompson: "God . . . divided the waters which were under the firmament from the waters which were above . . . The human body is a recapitulation of this . . . for the body itself is a firmament which divides the waters of the brain from the waters of the genitals. Because of the sacred numinosity of the waters, all fluids of the human body—saliva, sweat, semen, and blood are sacred and mysterious substances." *The Time Falling Bodies Take to Light*, 18.
5. Felicitas Goodman, *How About Demons?* 111, 116.
6. Mark 5:7–13.
7. Galatians 2:20.
8. "Twenty Minutes of Reality."
9. John G. Neihardt, *Black Elk Speaks*. The Grandfathers are invoked in the dedication of the account and throughout.
10. *Junky*, 22.
11. 126.

Chapter Eight

Smoking As Ritual,

Smoking As Addiction

And the Lord God formed man of the dust of the ground, and breathed into his nostrils the breath of life.

—GENESIS 2:7

Along with my wife and two children I had that day toured the curious city of Avignon in southern France. During fourteenth-century unrest in Rome, the popes had moved their seat here. Their palace still stood in good repair: a thickly walled and heavily fortified unit of buildings, replete with crenellated watch towers and a cathedral spire. The pope's bedchamber, we saw, was an irregularly shaped, smallish room, the five-foot-thick walls painted with a myriad of dainty pastel flowers, and morticed inward at two openings to allow light to enter through slits of windows.

When Gregory XI returned himself and his seat to Rome toward the end of the century, other authorities refused to follow, and the unthinkable happened. Suddenly there were two popes (or, from the point of view of church historians, one pope and an antipope). Although the sentence does not parse, there were two pontifex maximus—*two ultimate mediators between the human and divinely transcendent realms, two ultimate bridge-builders (*pont, bridge; pontifex, *bridge builder). In a grimly playful interlude, three persons claimed the papal prerogatives.*

The icy Rhone River descends from the Alps and sends its swirling steel-colored waters quite close to the papal palace. The river's power overwhelmed the engineering ability and ambition even of the Romans, but it is said that a bridge was later built by a Christian, Saint Benezet. Today the arches of the bridge step out from the palace side of the river, but stop short of mid-channel, aborted, a particularly violent head of waters having done in the arches that once stood there. Thus in the middle of the Rhone River, an horizon, a boundary.

That night we had dinner in a crowded restaurant in the commercial center of the small city. We were jammed in, with me facing the wall. At one point the tobacco smoke was so thick I turned around to see where it was coming from. I brushed against a young man and looked directly over his shoulder at the dimly outlined face of his tablemate, a young woman. Through the smoke she exhaled, her face emerged. Her wide-open eyes turned to slivers and dissolved into his, and her mouth stayed ajar tremulously. She appeared as would a goddess in epiphany.

I was interrupting. Quickly I turned back to my family. But for several moments the trance held me.

Thus she appeared to me. Did she feel herself to be a goddess in epiphany? How could we ever know? Even if she did, how could she acknowledge it—so weird and subliminal it would have to seem to her?

But why not plain facts? Her smoking may involve a mere nicotine addiction. Moreover, if habitual, it may lead to her premature death, as scientists are ever better able to explain. But physiologists' and physicians' cold and plain facts are only the most obvious features of what we do; they are necessary but insufficient to explain us.

It is the larger domain of human significance, and the fantastic things we often do to promote our sense of it, that engages us here. There is a deep fascination in activities like smoking. These more telling facets of smoking become plain once we take seriously the concept of body-self, and refuse to think of self's body exclusively as scientifically focused and manipulated object.

Communicative Ecstasies of Body-Self Forming Itself

The case at hand is just one example of the ritualistic possibilities of smoking for body-self to form itself. The ritual may be rooted in traditional origin myth and reliably regenerative. Or, more likely in this case, fragmented, shallowly rooted, with questionable consequences.

We can say of smoking in general: the serpentine curl of smoke exhaled from the inner cavities of one's body negotiates the boundary between the private inner cavities of body-self and the public world. It is communication that expresses and validates in the public world that the private wilderness self remains real and potent. What it is exactly is not known. That the privacy is—that is certain. The smoke crosses horizons and boundaries so that body-self in its inner private aspect and in its public one are both demarcated and connected—body-self defined and reconstituted in one way or another. Body-self shares itself and returns to itself characterized through others' recognitions of it. It is numinous ritual that ecstatically builds the self, either reliably and regeneratively, or merely momentarily and deceptively.

Engorging smoke and then expelling it, we send it out, shaped, from the inner body-self. We present ourselves in the nebulosity before witnessing eyes. Ours is the sensuous sense, funded in the bodily memory of the human race of what recognitions happened "in those days" on Earth among our vast kin, natural and supernatural, human and animal. It is no accident that in Christian ceremonies the Holy Spirit is sometimes named the Holy Ghost. The smoker is mimetically engulfed—constructively or ephemerally—with supernatural or mythic presences, those originating totemic powers that animate all religions, no matter how cold and sophisticated their institutional surfaces finally become. The invisible is made visible.

The supreme divinity is often conceived as *causa sui*, cause of itself. Expelling smoke and emerging through it as if in epiphany, we mimic this supreme power however erratically or dimly. We buy or roll the cigarette, place it in our mouth, destroy its substance as it is transmuted by fire and the smoke is inhaled. The smoke emerges again with the reality and mysterious message—even with a hint of the volume—of one's private bodily self as it crosses into public reality. The self creates and recreates itself in the world's interfusion and excitement.

We are essentially private and essentially public, and essentially animal and essentially strange in our human animality. I think we long to know other body-selves' experience as they engage personally with the world. At least equally poignantly, we long to be known by other humans—at least sometimes—as we ourselves engage immediately in the world. I think we also always want and need to be known by nonhuman others, particularly when we are very alone and forsaken by humans. (Recall Saint Paul's reference to heaven: "Then we will know even as we are known."[1]) By smoking, our privacy and bodily reality is recognized

without being violated (at least not obviously) and we swell magnificently, if only momentarily, enjoying ecstatic experience: being a self.

Inner Body As "Womb"

Smokers want to plant their body-self in the world's community and to
establish their significance: the ingestion and regurgitation of smoke
signals generative and regenerative capacities and contributions. As if
the inner body were a perpetually fruitful womb that participates in the
womb of Nature herself. As if the apparition issuing from the cavity of
the body were a second birth of self, one charged and amplified
through our own and others' confirmation of it. A development and
recapitulation of the first birth? Freudian psychiatrists' reduction of
smoking to fixated orality limits attention to dependence on mother
and masks out Mother Earth floating in her filmy membrane of air.

As we will see more clearly in the next chapter, boundaries between
private and public and here and there cannot be crossed powerfully
without blurring the lines that divide present from past, and future and
possibility from actuality. Certain levels and sorts of trance are induced.
These deeply personal communications and encounters are simultaneously super-personal and anonymous, in an important sense mythic.
They have their source in an experiential realm felt to be numinously
and unconditionally valuable that cannot be precisely measured and
located.

John Dryden's fugitive lines suggest smoke's mythic reality, the eternal present:

> *Nothing dies;*
> *And here and there the embodied spirit flies.*[2]

Fragmented Mythic Impulsions

The young woman in the restaurant evoked this mythic realm, at least
momentarily. Engulfing herself for a few moments in a stray goddess
in epiphany—if it was that—is far from transferring her infantile
mother-need into an intelligently informed reverence for Mother
Earth. It cannot substitute for initiation in a culture's rite of passage
and subsequent maintenance and work. Smoking, she may be merely
posing as an adult for the moment.

In her afflatus she may have leapt into a situation trembling with
explosive possibility. Perhaps her tablemate is a compulsive and imma-
ture man who, along with her, could no more raise to maturity potential
offspring of their union than they could reach it themselves.

Fragmented and shallowly rooted, riteless, contemporary secular and scientistic culture presents itself naked, almost totally out of sync with the regenerative rhythms of Nature. It is lacking especially the generational rhythms that induct us into sound individuation at the various pivotal stages of our lives. Paul Shepard writes:

> Like land birds instinctively setting out on transoceanic migration, given assurance . . . from the experience of the species that there is land on the other side, the human adolescent organism reenters the dangerous ground of immature perception on the premise that society is prepared to meet his psychic demands for a new landing—that is, that society is organized to take these refractory youths through a powerful, tightly structured gestation; to test, teach, reveal; to offer . . . things worthy of their skill; to tutor their suffering and dreaming; and to guide their feelings of fidelity. If . . . the adult group is not prepared to administer the new and final birth, then the youths create autistic solutions to their own needs and . . . sink finally, cynically, back into their own incompetent immaturity, like exhausted birds going down at sea.[3]

These adolescents never get beyond the stage of dependency on mother—perhaps blaming her for all and sundry—and into that of Mother Nature: the world itself to which we must adapt, and which it is pointless to blame.

Achieving Sound Individuation

Sound individuation of the mature body-self is possible only because we can collect or center ourselves, dampening others' presence within us, whenever it is appropriate or necessary to do so. But whether collected or not, we have an implicit, wordless awareness of the privacy of our inner bodies, although in engulfed moments when we little fear violation this awareness is not focal. Constant and contrastless, privacy is startled into explicit awareness when violated some way, as when an insect flies into nose or mouth, and, revulsed, we expel it. The sound body-self experiences both a closed or unique and a communal or shared reality, with different emphases at different times. The demonic self is terribly violated.

Being caught talking to ourselves in the presence of others is embarrassing, because we all implicitly believe that privacy is each individual's sacred right to possess and each's obligation to guard. Unshielded by ritual, even theorizing about these matters is slightly embarrassing.

The self as an individual body caught up in awareness evolves delicately and very vulnerably, since authoritative others can violate it by denying its right to privacy. This happens most brutally when entry into the cavities of the body is made without having been responsibly invited.[4] It happens even when a person incorporates the belief of others that nothing can be kept hidden from them. Victims of demonic possession have been made particularly vulnerable by such abuse. Terrified that the contrast between self and intimate other might collapse, they despairingly attempt to eliminate the possibility by incorporating the threatening other and making it their own; but they become possessed. This is a panicked response, apparently irrational, and yet the only way open perhaps for that individual to continue to be at that moment. Since we are ecstatic creatures who must at crucial moments maintain the clear contrast between self and other, the attempt fails wretchedly.

An individual is typically possessed by the other who did the violating. But perhaps the rupture or puncture of self leaves an opening for addictively fixated substitutes for the original violator, ones that mask his or her transgression.

Grandmother Wilshire, Mary Bittenbender, was born in 1859 and raised in harsh circumstances in western Pennsylvania. Her only child, my father, was born when she was forty-two. I remember her in Los Angeles in the thirties— slipping around the house, on the margins, in perpetual shadow, frail, taciturn, relentless. I would sometimes be consigned to bed with Grandma when I would awaken in the morning before my parents and become noisy. Sometimes her quavering voice would bring me stories. One was of herself as a child welcoming the boys home after the Civil War. Another was of herself as a teenager quarantined with her younger sisters in a barn during a diphtheria epidemic one winter in Pennsylvania. She would reach her fingers into their throats to extract the phlegm.

When I was about eight or nine, mother told a story about my father and his mother long ago. Grandma suspected he had broken a neighbor's window playing ball. He refused to tell. One night while he slept she shook him awake and demanded, Did you break that window? He confessed.

I remember you, Grandma, and wonder about you.

Communication of Privacy
Privacy exists, but embedded within a public matrix, with ritual ways of signaling this relationship. In good communities each person recognizes everyone, including oneself, as recognizable on the basis of public

display, and also as private. When privacy is revealed (though always partially) in inappropriate circumstances, human communication and human being rot.

If "spirit" and "spirituality" did not inevitably connote what is other than body, I would call the body's manifold powers of communication its spirituality. They vividly proclaim a sense of self—as self to be confirmed by others as private and public. Without this sense I do not think we would be self-conscious, would *be* human selves; we would not be significantly different from other animals.

I do not want to give up completely the force of "spirituality," but must completely reconstruct the context in which it gets its meaning. Whenever the term is used henceforth, I will attach two asterisks— **spiritual—to remind us of its special use. The body is **spiritual in its powers of communication, in its abilities to express, recognize, be recognized, and to incorporate others' recognizings. The body is **spiritual in its dangerous interfusions and cycling transactions in the world. To be sound, it must both express and preserve its privacy in a public world. The **spiritual body is body-self, able to keep its own council if mature, able to take initiatives and to acknowledge its own guilt if indeed it is guilty, and able in the most extreme situations to accept its own death.

Each group or corporate individual generates its own controls of communication. These are its ritualistic base that tie individuals into groups, and tie both at a present time into the past and future. The wisdom of indigenous or traditional peoples requires that all human control occur within will-of-the-place, the wilderness whole of Nature. (Again, *Nature* and related terms with mythic force are capitalized because generally felt to be unconditioned and unconditionally valuable.)

Each group develops its own religion and folkways. These are the literal ways that the folk, through ritual communications in bodily media, regulates intercourse across the boundaries between civilization and wilderness, human and animal, private and public, now and then, life and death. Traditionally, crucial transitions in individual and group life have been interwoven with Nature's transitions over the seasons of the year. The role of animals in human life was regulated and confirmed. Cultures reenacted implicitly versions of Aboriginals' Dreamtime in which humans were perpetually regrounded in their Sources. If sound development of body-self requires a ritualistic base, and I think it does, we live in extraordinarily dangerous times, for we are nearly riteless.

Consider the case of feral children. Some are believed to have had some of their primal needs satisfied by female animals. Rediscovered by the human community, they exhibit some of the characteristics of their first role models—a four legged gait, barking, grunting, etc. Given human adaptive and mimetic capacities this is understandable.

Are these beings human? According to biological criteria they are, but these criteria are not sufficient here, for they cannot alone guide our evaluations and decisions. Something is awry in the **spirituality of the bodies of such beings. They lack recognition as humans, and this can be supplied only by humans.

Urbanites today suffer the inverse of this condition. With the general unraveling of ritualistic life, we tend to get too little confirmation as organic beings from nonhuman sources. Animals, birds, and fish recognize our existence. These creatures will not *say*, even subvocally, "human" or "animal" when they encounter us, but they recognize us in their own terms. To lose their recognition is to lose our cycling communing and exchanging with the wide world, losing our substance. As Gary Snyder writes, "The Buddhist iconographers hide a little animal face in the hair of the human to remind us that we see with archetypal wilderness eyes as well."[5] I need not confront a wild animal and be fixed in its gaze, but can empathize with it so that its gazing intermingles with and reinforces mine.

Confirming the Private Body-Self

As I have attempted to show, our experiencing is not hermetically sealed from other things' experiencing. Minds and selves are not separate, immaterial entities inside anything. The most delicate, intricate communication must occur between them. Only when we regard ourselves abstractly as mere homogeneous objects or units do we feel ourselves separated or isolated from others. If we succeeded in destroying all wild things, the wild, half-lit parts of ourselves would no longer be confirmed by them. Hidden inner organs—particularly the heart and viscera that generate the great emotions—would lose resonance with wilderness, lose exuberant vitality. Our privacy must be expressed if it is to be fully constituted and confirmed. But central to the expression is the expression of incompleteness of expression. This humans appreciate—and perhaps animals do as well!

But certainly any recognition of us by animals confirms a basic level of our being in which we are mute—or in which we communicate through sounds alone. Our writable and speakable language is powerful and unique. But thinking that that is the essence of us leads to addictive and desperate loquacity. We need animals' wordless rapport. Even terror in the face of wild animals (though not hysteria) strengthens us. It confirms the wild parts of ourselves, as Paul Shepard notes.

In some obscure but fundamental sense even vegetables and trees recognize us, and their recognizing confirms our private vegetative processes in a mysteriously erotic way. Glen Mazis writes:

> The deepest source of erotic energy is that of perception itself: the entire sensual realm and all its objects are through their communion in perception in some sort of energizing intercourse . . . rock, bark, cloud, and grass speaking in erotic whispers . . .[6]

Also recall Emerson's observations about boughs of trees nodding to him familiarly. We share with plants nutritional powers: we ingest, digest, excrete, and along with them we soak up the sun and moisture and need the Earth. Scientists now seek to confirm an hypothesis they think probable: the very cells of *all* living things, evolving over untold millions of years on the rotating earth, have built-in, night/day "biological clocks" that massively influence all the body's functions.[7] To the extent that our "Nature transcending" twenty-four-hour lives ride roughshod over our own cells, we are less, not more, human.

Our encompassing set of kin includes even inanimate things such as mountains—vital presences in us, since on some primal level we feel ourselves seen and felt from "their point of view." However speechless we and they may be, we feel their recognizing of us as this flows into our recognizing of ourselves. Perhaps we interweave dancingly. We cannot directly control this encounter, nor should we want to. Our place in the abiding matrix of Nature is confirmed, so *we* are.

In fact, the sundering of human recognition from animal recognition in particular, and the general suppression of animals, are yet further instances of the dualizing lines drawn around ourselves in North Atlantic culture, isolating us in the sterile objectifications of scientism and secularism. Our privacy belongs to wilderness and wildness and can never be fully known to others. But *this* must somehow be communicated. Obtuseness here breeds loneliness and addiction—wizened fruits of superficial civilization.

Some in North Atlantic culture begin to see the need of relationships across the whole domain of Nature. Reports circulate of the therapeutic role of animals in nursing homes. Several years ago an article appeared about wild animals—coyotes were some of them—that New York apartment-dwellers had sequestered in cages or in whole rooms of their own dwellings.

The common response—Crazy! Such was foretold by Henrik Ibsen a century ago in *The Wild Duck*. The "enlightened" man discovers the wild creature in the simulated wilderness in his friends' attic and ridicules and destroys it and them, everything.

Nowhere does the wisdom of indigenous peoples' modes of existence and communication better show itself than in their communication with or about animals. Harold Searles believes that we (presumably we moderns) both want and do not want to be animals.[8] "Primitive," ritualized modes of communication make this ambivalence more than tolerable. Ritualized smoking, say, expresses and constitutes us both as breathing, mortal, particular animal bodies, and as makers and users of sign and symbol systems, the manifest elaborateness of which distinguishes us from the other animals. Jamake Highwater:

> At the conclusion of the pipe ceremony the participants murmur: "We are all related." The act of smoking is a ritual communion with everything in creation, with every possibility of being—what lies before us and also what lies beyond our understanding and knowledge. "We are all related."[9]

This means all beings who emerge and pass away in the regenerative cycles of Nature. Typically today, smoking amounts to perversion of regenerative celebration.

Orgasmic Transactions

When we keep the category of religious experience very wide and elastic, as Emerson and James advised, we better understand ecstatic experience. If our deepest need is to be, nevertheless it is not always to be something precisely specifiable here and now. We need our primal, ever incompletely knowable private and ancient wilderness selves confirmed, our serpentine liminal bodies crossing boundaries honored.

Don't we need sometimes to lose our conventional orientations and boundaries in space, time, and culture, and to be periodically possessed angelically in some degree? I needed my dog. James wrote that we need "moral holidays."[10] Not that we should flout moral duties, he thought, but we need and want those protected times in which we safely put down our guard and enjoy the dynamically coordinated inner and outer reaches of ecstatic abandonment; enjoy that orgiastic dancing that blurs and melts partitions. Most emphatically during sexual orgasm, we allow the wild body's own will-of-the-place to move through it, and us, convulsively. As in the belly laugh, but more intensely, the involuntary aspects of body-self extend and enliven the rest of it. Fullness of self is celebrated, a fullness more powerfully overflowing than any verbal or mathematical individuations of its aspects or functions could ever suggest.

The orgasmic demobilization of routine behavior is, at the same time, a disintegration of the linear timeline. Orgasm: life as blind, freefloating, as blissful and careless as in the womb. Yet it does not occur in the womb. It occurs in places today that are typically very unsafe and unprotected by ritual.

Various distinctively human orgasmic experiences seem to be necessary to restore our balance as creatures who never cease to be animals—bodies located in some local environment—though multidimensionally ecstatic and strange ones. Orgasmic rhythms, perhaps belly laughs, are needed to reweave us in the natural world's—the will-of-the-place's—irruptive events and regenerative periodic cyclings, both. Relief: for the moment we are appropriately out of control! Such times seem necessary for body-self to appropriate all of itself as "I myself—me."

But outside ritualized sequences that are interwoven with the Earth's regenerative cycles and our basal moods, orgasmic rhythms are not sufficient. Their very frequency and irruptiveness may lead to desperation.

Sadly, tobacco smoking in urban cultures is not explicitly ritualized and easily degenerates into mindless mechanical addiction. Instead of opening the way for **spiritual growth, it causes body-self to shrink and crumple within itself. Body-self is at odds with itself: the addiction's grip on the smoker is proportional to its ecstatic inadequacy. Ecstasy at its best is standing open to ever more growth, ever more cycling reciprocity that moves across horizons into the unimagined. This weaves us whole. Mere nicotine addictions sell us short.

A group of colleagues had piled into a car to drive somewhere for lunch. I was last and sat squeezed in among several others on the back seat. An acquaintance had lit a cigarette which was beginning to upset my stomach. I started to ask him to roll down the window. But he immediately opened it and threw out the cigarette. I said something like, "I'm sorry to spoil your pleasure." He replied decisively, "It's not pleasure, it's smoking."

More Mythic Dimensions of Smoking

Smoking involves destruction, and this figures in who we are. Typically the supreme divinity is conceived to be supreme Destroyer as well as

Creator. In smoking the fire is birthed and dies then birthed again: a faint echo of the most ancient of all cultural-biological-religious themes—birth, death, rebirth.

In the immensity of space and time and the all too frequent **spiritual distance that divides us from other body-selves, there is nevertheless this little light and fire to call our own in the midst of this immensity; nay, this little light and fire to keep ourselves burning—ourselves as alone *and* public, bodily *and* **spiritual-bodily.

Now we know that smoking destroys not only cigarettes, but also often the human body sooner or later. That many still smoke suggests how strong is the need to establish identity by achieving recognition of both our private and public aspects. Many are willing to destroy the very body that is recognized to be one's self. The body's serpentine kundalini energies turn upon themselves and stub themselves out in a desperate attempt to affirm and recreate the self. If dying some time is the cost of being now, so be it.

Usually, the progress of addiction is dull deterioration. The body's identity building energies fall into substitute channels involving death of self without rebirth, and finally stagnate and putrify.

It's all very complex—often vexing and dispiriting. Is it any wonder that many secretly or openly want to die?

Many years ago—I must have been about twenty—I was driving with my parents in the mountains of California. We stopped at a turn-off to enjoy the view. Without thinking I lit a cigarette. Mother said, "Why do you do that?" She objected mainly, I think, because it branded me a member of an alien group. Almost as automatically as I had lit it, I blurted, "Because I need to." She had no response to that.

I was completely and strangely elated and surprised, and have been trying since then to understand what I meant. Who am I? Who is she?

Self Reconceiving Itself

In traditional dualistic views, we are always centered because the soul or mind is somehow inside the body, existing there in one point instant after another. If the self is inside the container-body, it is separated from other selves inside their containers. This crudely spatial view cannot fathom human intercourse. It cannot grasp our body-self reality as we move multidimensionally in space-time-culture.

Essential for soundness of individual selves is continuity of body memory. Our sense of smell in particular—as we shall see—enlists the sustaining reality of time lived by us bodies. Primal communications with others of all sorts, alive now or in the past, weave us into coherence. Body-selves feed into each other.[11] Addictions are pathetic stop-gaps.

Even a little understanding of everyday mythic-organic-magical reality is satisfying—as Michel de Montaigne appreciated: "Truly man is a marvelously vain, fickle, and unstable creature, on whom it is difficult to found a certain and uniform judgment."[12]

In 1973–74 our family spent a year in Europe. The parents of the family who rented our house for the year agreed to take care of our dog, Buppie. When we returned the following spring, I saw that his pen was empty. "We could not control the dog and had to take him to SPCA," they explained.

My recollection of the event includes no memory of a response. I must have swept aside what had happened and covered it with busy-ness.

When Calvin Martin read a draft of this book and inquired about the dog, I told him the story. He said, "They killed your dog." When I made little or no response, he continued, "They killed your dog. You know that, don't you?"

"I feel badly enough about that already, Calvin." That kept him quiet. It did at least that.

Notes

1. I Corinthians, 13:12.
2. Quoted in *Selections from Ovid*, ed. F. W. Kelsey, Boston, 1890, 175. I cannot locate the source in Dryden.
3. *Nature and Madness*, 65–66.
4. Correlation between multiple personality disorder and sexual abuse in early childhood is high. See Colin Ross, M.D., *Multiple Personality Disorder*: "The MPD patient has fragmented self caused by real physical and sexual assault." "Most MPD patients are victims of violation of the incest taboo," 10, 13.
5. *The Practice of the Wild*, 20.
6. *Trickster, Magician, Grieving Man*, 183.
7. *New York Times* Sunday Magazine, January 5, 1997, 48. Also *Science News*, December 6, 1997, 365.
8. *The Nonhuman Environment*, 375, 420. This is an unusually expansive view of human relationships, given the date of its publication (1960), and the fact that the author is a psychiatrist.
9. *The Primal Mind*, 189.
10. See, for example, the last page of "On a Certain Blindness in Human Beings," my anthology, *The Essential Writings of William James*.
11. Christiane Northrup, M.D., in her newsletter (October, 1997): "Think of

your social network as your immune system safety net. At some level, every human illness is connected with lack of social support." She cites a study in *The Journal of the American Medical Association*: cold virus deposited in the noses and throats of those with wide social networks were less apt to get a cold than those with narrow (2). Perhaps the common cold is not such a mystery after all.

12. "By Various Means We Arrive at the Same End," *The Essays of Montaigne*, chapter 1. For mythic and magical reality see Jean Gebser's monumental work, *The Ever Present Origin*.

Chapter Nine

Body, Nose, Viscera, Earth

If the myths record the beginnings of things, then every recitation of them in ritual becomes coeval with the primordial past . . . knowledge of the passage of time . . . is evaded, and the world is recreated anew on the original model.
—FREDERICK W. TURNER AFTER MIRCEA ELIADE, *BEYOND GEOGRAPHY*

The very conception of a beginning of conscious life carries with it a paradoxical reference to something prior to that beginning . . . As we examine our own duration in time . . . we can find no wall of partition between self and prior-to-self. I never know by introspection how old I am, or that I have a finite age. If the impulse which is I is a "racial impulse," there is no reason to ascribe age to it: it is presumably, like energy, always new as on the first day . . . It is, in fact, a gratuitous assumption that where one self begins another must stop . . . different selves overlap. . . .
—WILLIAM ERNEST HOCKING, *THE SELF: ITS BODY AND FREEDOM*

I had been swimming in the large university pool and had lost track of time. Seeing I would be a bit late for an appointment, I was hurrying now out of the building toward my car parked a little distance away in the lot. Though I'd showered copiously, I still carried an aura of chlorine, my damp towel thrown over my shoulder and gym bag in hand. I walked as fast as I could. The most

direct route to the car lay through a thin margin of trees. I had nearly passed through when it stopped me. That smell again! Not the chlorine I carried with me, but the scent of pine from decades ago in the Sierras. That smell was back again, and I was that boy once more.

Looking around I saw I had walked through some pines bordering the parking lot: their boughs bent gently in the breeze, nodding familiarly. I put down my bag. There was much more time and space for me than I had thought.

Usually only the smallest part of reality is focused. The rest is lost in dim margins of awareness or absent entirely, though presumably the universe works on each of us beyond anyone's ability to imagine. Even on those uncommon occasions when I become aware of the Whole, I have only a very incomplete sketch of what is or might be going on, though the awareness greatly refreshes.

As we navigate day by day, things must sometimes be questioned. But to maintain our balance requires that we leave most things unquestioned. Reality lived is the intuition and feel of it all circulating, pressing, releasing through us. Without blind trusting on some level, we cannot stand open to the world as it flows into us, circulates within us, and flows out—with consequences only fractionally imaginable. There is no way to accurately calculate output from input, response from stimulus, and to try to do so would eviscerate our ever present spontaneous powers. Without trust, we do not stand open to our primal needs and hungers as body-selves formed over millions of years in wilderness. Substitute cravings take over addictively, substitutes mockingly "me and not me." Stupified by the dissonance, we mechanically repeat ephemeral gratifications.

Whether manifestly religious or scornfully irreligious, we cannot live soundly without a visceral sense of unconditioned, total reality, that is, a mythic commitment to "the everything else," an ill-defined numinous presence and power. It need not be acknowledged, and we need not be sound. Disoriented, impelled by momentary impulses, we can plunge and crash about.

Body-self is mature when it knows viscerally that it is both a unique center of agency *and* bound indissoluably into the unimaginably vast, regenerating world. Genuine ecstasy allows the mystery of growth. J. Krishnamurti captures it in his title, *Freedom from the Known*. Josiah Royce has his own angle on growth as spontaneous play:

> As the infant that studies its fist in the field of vision does not know as yet why this curiosity about space and about its own movements will be of

service to it, so throughout life there is something unpractical, wayward, if so you choose to call it, in all our curious questionings concerning our world. The value of higher insight is seldom immediate . . . [There is] an element of noble play about it.[1]

Any ecstasy not associated with growth is addictive and degenerative.

The Dimensions of Time

Maturity means we reclaim our bodies' propensities and skills built up over untold millennia as we adapted within Nature. Means, that is, that we reach behind the relatively sedentary life of agriculture: that grasping at a bit of earth to own, to pen ourselves up in, that occurred a mere eight or nine thousand years ago. We *are* self-moving bodies in space and time. We needn't walk only in order to get somewhere, but may walk just in order to celebrate and confirm what we are: walkers, spontaneous movers in space and time. Thereby we feel "I-myself" through and through, which is exhilarating.

Thoreau writes early in *Walking*:

> I have met with but one or two persons in the course of my life who understood the art of Walking, that is, of taking walks—who had a genius, so to speak, for *sauntering*, which word is beautifully derived "from idle people who roved about the country, in the Middle Ages, and asked charity, under pretense of going *a la Sainte Terre*," to the Holy Land, till the children exclaimed, "There goes a Sainte-Terrer," a Saunterer, a Holy-Lander. They who never go to the Holy Land in their walks, as they pretend, are indeed mere idlers and vagabonds, but they who do go there are saunterers in the good sense, such as I mean.

To be sauntering—exploring, observing, dreaming, open to the unexpectable—is to be already in the Holy Land. It is the Holy Land. Celebrating itself, body-self finds itself intrinsically valuable, vital, coherent and continuous. Don't neural and muscular habits of prehuman and human hunter-gatherer forebears come alive? Then homecoming: humming awareness of belonging, of wholeness and haleness, of holiness?

How do we live and know space and time viscerally, when they are more than the abstractions, "Big, big box," and "Long, long clicking rails"? Most fundamentally, time is not limited to one dimension: steps, stages, moments that come and go and are left behind to vanish. Time is not just linear but cyclical, a perpetual dipping into Sources, our ever present origins, an eternal present. And space as we immediately experience it and live it is not an infinitely large box in which we rattle around—not if we are sound. It is spatio-temporal, at one with our experience of movement and time, at one even with the eternal present.

The Most Intimate Horizons

Not all horizons can be seen. The horizon that divides the inside from the outside of body-self, its privacy from its publicity, can be "seen" only in an equivocal sense. If one looks inside one's mouth with a mirror—or into someone else's—only what is in public space is seen. The "ownness" of one's own immediately lived privacy is experienceable as such only kinesthetically and kinetically by body-self itself; it can never be seen by anybody. In "Kubla Kahn" Coleridge intuited the fertile darkness of the Earth and our earthen bodies with their hidden cavities, organs, fluids:

> *Where Alph, the sacred river, ran*
> *Through caverns measureless to man*
> *Down to a sunless sea*

The body's immediate experiencing of its own dark insides is prereflective and nonobjectifying, forming the vital root of "I" and "I-myself." Each can grasp himself or herself as more than can be grasped, as what forever overflows calculation and description. Responsive to these depths, we are primally responsible; we know that others can experience the same sort of thing. We speak of this responsiveness of body-self as the **spirituality of the body—asterisks added to jar us from the dualistic assumption that spirituality cannot be bodily.

Likewise, the horizons of time, the moving boundaries of *now* that divide present from past and future, can be "seen" in only the obliquest sense. But we live these horizons immediately, and sensing them are inexorably beyond them in the indefinitely past past and indefinitely future future. As William Hocking points out, even if we try to imagine a beginning of time, we must immediately feel the "other side" of this beginning, an immemorial and primordial past. The young woman emerging through her cloud of smoke may be one with goddesses and gods in mythic consciousness, in a kind of eternal present. She communicates serpentlike across boundaries and transcends the atom of time called the present. This is her **spirituality—her contact with archaic presences that indwell and constitute her ecstatically—at least for the moment. Our modern departure from ritual moves us into uncharted waters.

I smell that smell of tobacco! All the past epiphanies of faces materializing through their clouds present themselves again, relived in that peculiar way in which time turns back into itself, recuperating itself, flowing but also cancelling itself—time is bi-dimensional. The young woman emerging through her cloud is one within ghostly legions of others emerging through theirs. With extraordinary power, smell vibrates ecstatically across the horizons of time, fetching what lies beyond as no other sense modality can, returning us home to our primal body-selves. If more than merely desultory, smell carries us through black holes in time. We needn't clutch for satisfaction of cravings *now*.

Finding Our Way Back into Olfactory Experience
With the exception of the kinesthetic sense—the immediate feeling of our bodies gathering and conserving themselves through time and providing the fluently moving "platform" upon which all the other senses operate—with this exception, the olfactory is most overlooked and undervalued. It is least used by science. The abstractions within which scientific data are conceived and the instruments used to detect the data typically accord no relevance to smell. Such data cannot be "hard," cannot function within our typical information processing systems. But this only shows the limitations of these systems. For surely it is through smell that we vividly know foundational regenerative features of our lives across horizons.

But more and more we rely upon vision to find our way around in a technologically structured world, and more and more lose touch with olfaction. Vision is the distancing sense *par excellence*, because it so accurately discriminates and divides, separating us from the other animals and placing us in what can seem to be a grandly isolated present moment. Peering at animals through binoculars we may mutter, "See how different we are." But we all do have noses. Could one reason for neglecting the sense of smell be that we associate it with animals? Recoiling at a dog's sniffing our private parts, do we seek to deny our common animality? Culturally and ritualistically enforced protections of privacy are necesary, of course, but are stultifying if prompting isolation.

Is this similar to our shunning alien humans because they suggest bodily features of our animality we would rather forget? But our affinities with animals will not cause panic if we keep attention fixed on the unique **spirituality of the human body. We are both deeply kin to other animals and distinctive.

Vision distances and discriminates more sharply than does any other sense modality. Its current alliance with nearly ubiquitous commercial and technical enterprises eclipses our whole bodies, our ambulatory mobility, our olfaction, and so on. Modern developments eccentrically emphasize perennial needs to establish identity by distinguishing our bodies from others'. "I must get my food into my mouth, him to court at 9 a.m., you to the airport tomorrow noon, that rat into that maze, and so on."

It's true, of course, that practical life has always demanded that I as an adult be capable of orienting myself within myself regardless of that to which I am attending. "That's to the right of me, this other thing to my left. I can turn as I choose. I am distinct from everything else." I must be clearly set off from all other beings, and as each moment passes through the movable vital mass that I am, each recedes farther and farther into the past.

Scientific inquiry, technical and commercial enterprises, and practical coping are all based on linear time. But there is more to our lives than this! The ways things divide and fall apart does not exhaust reality, and sometimes distracts from it, weakens our momentum through time. First, the presence of others animates my own presence to myself, for I tend to absorb their attitudes toward me as my own attitudes toward myself. In primal experience even stones appear as friendly or alien presences, and I appear as companionable or not. Second, I do not just leave the past farther and farther behind as I live in the present. When caught up excitedly in other things or persons, the line dividing self from other tends to blur, sometimes approaching, at times reaching, the vanishing point. But when the body-self's experien*cings* blur into the world experien*ced*, they also tend to blur into each other. That is, the present and the past blur into each other.

Why? When I no longer experience myself as a subject vividly set off in this moment from others as objects, my sense of here blurs into my sense of there. But on the primal level of ambulatory experience a distance in space is also a distance in time, a lapse of time. If we were to focus our consciousness, we would say, "We're two hours' hiking away from the lake." But during primal moments of experience we do not focus it and objectify space and time. Still, there is an implicit lapse of time involved in getting around in space, and that lapse does itself lapse. The Here and Now blurs into the There and Then.[2]

The lapse of time is not grasped clearly as a lapse; the lapse itself lapses; there is melding, blurring, of past and present. This is not mere confusion. It is that normal, vital momentum in which life is lived eagerly and well.

A Pivotal Point

I will go through this in a different way. If we do not understand mergings we will not grasp supportive and buoying mergings with compatible others, won't appropriate what remains of our mythic roots. Nor will we understand mergings with incompatible others that infest and fix us in addictive isolation from ecstatic resources, lodging us in black holes in time.

As an experiencing-experienced being, body-self can contrast its experiencing of space-and-time to everything else that is merely experienced or experienceable by it. But only when we see how momentary and at times precarious this construction of individual self is, can we understand how personal individuation is debordered at crucial times and caught up in others, and in corporate individuals of all sorts—potent quasi-mythic groups, say. The past abides, and this is no mere "in the head" recollection. Olfaction is no superfluous primitive sense we can lose with impunity.

The body-self knows the world because it can face things and handle them—literally or figuratively. In facing space and in advancing into it, the body-self also faces time—the future—and advances into it. In anticipating something and then responding to the fulfillment of one's anticipation, one passes it. It is left in the past as something filled out, set, determinate.

So far everything seems clear. The perceiving being, "the subject," emerges in contrast to perceived things, "objects," and the future, the present, the past are distinguished. This holds notwithstanding our ability to shift from perceptual to imaginative modes of experiencing: to float around, so to speak, in space and time, to escape the iron rails of linear progressions. But for the sane, imagining is only imagining.

Notice, however: the ability to make these distinctions—between imagining and perceiving, self and other, here and there, now and then, sane and insane—presupposes a calm, somewhat detached, highly individuated being. And that is just what should be questioned. At crucial moments—and for more than moments—we are caught up powerfully, moodily, in the other (of whatever sort, whether past or present). As the distinction between self and other blurs, the memory of one's own passage through the world as a perceiv*ing* being loses its contrast to space-and-time-perceiv*ed* and blurs into it. In other words, as lapses of time between objects perceived in space do themselves lapse, the articulated world of concrete perceivable things in their own places and times blurs in the direction of a merely imagined world—a world of mere possibilities.

But, of course, it is not a clearly imagined world either. For distinctions between imagining and perceiving, possibility and actuality, here and there, now and then, and self/other blur. Resulting are trancelike states of possession—demonic or angelic—usually subtle, sometimes obvious. Noticing these states may be difficult.

Our actual human history and behavior can only be explained if we suppose that we human beings are frequently "not in our right mind." Without supposing this blurring of distinctions, we cannot begin to explain the precariousness of human identity. We cannot explain widespread failure of responsibility, and the fearful defensive-aggressive treatment of each other and of Nature herself. Genocides, abuses of children, lynch mobs, St. Vitus-dancings, orgies—various forms of madness, divine or demonic—have been, and continue to be, salient components of our history. Vocabularies reduced to fit the technological world of simplistically conceived individuals, entities, factoids, most systems—"newspeak" as George Orwell put it—mask out realities with neurotic fear.

No wonder, then, that persons unequipped to deal with archaic fusions recoil in the face of body-self and Nature, and attempt to force a continuity and identity for themselves through mechanical repetitions, addictions. This is fear of ecstatic possibilities, a fear of fear itself that represses the possibility of holy fear, awe, restraint, respect for all beings—including ourselves.

Despite the dangers, blurring just may spread into the ever regenerating sources of our being, powerful mythic presences and allies still extant that spring us from black holes in time, recreating the present. The young woman emerging through her cloud of smoke is not necessarily caught in mechanical addiction. In bonding perhaps with a goddess in epiphany she emerges powerfully in the present moment. But is she oriented over the long term in the abiding, cycling earth and its presences? This question haunts.

John Dewey's Encounter with Body-Self and Time
How labile and spontaneous is the primal body-self, how various are its mergings, and how vastly different in value! We may absorb the pres-

ence of incompatible and condemning beings. Fed into us at an early age they cast a spell upon us, and wither potentialities.

This dangerous, labile level of self is what John Dewey discovered in momentous sessions with the therapist F. M. Alexander. Dewey had encountered the inability of the most penetrating thought to get to the bodily matrix that conditions thought and being.

At age fifty-seven and in the grip of a pervasive midlife crisis, Dewey betook himself to Alexander.[3] The "body-worker" (or perhaps psycho-bio-therapist would be better) concentrated on altering the way Dewey's head rested on his neck. This habitual positioning was so basic to his identity that he could not imagine new positionings until Alexander moved his body for him. When the heavy head is not balanced and poised atop the backbone, a dynamically integral part of the body, the most intimate continuity of self—experiencing's coursing in time—is knotted. Our whole sensing body is constrained by fixed ideas and mind games, unable to grasp new possibilities for growth. Instead of acknowledging origins and building upon them, we are captured by some of them. Head still tilting up at people who once loomed above him, teeth clenching, the middle-aged Dewey at some level was frozen in his past.

Rebalancing the head on neck and backbone may be sufficient to free the self for mature behavior, sufficient to allow the self to individuate itself as an adult at this moment because it acknowledges the whole past world as past. The past would not be "acted out," dribbled out. The actual momentum of time can be released only by the whole, aware, integrated body. To align and straighten out the body is to allow lived time to wash out panicked fixations, addictions. The past is not always present beneficially.

Such therapy may be sufficient to liberate awareness by opening it to some of its powers here and now in the body. No longer blindly attached to a complex of meaning and behavior laid down in the dim past and recurring mechanically, one is restored to regenerative cycles. The two dimensions of time, the linear and cyclical—or eternal present—are distinguished and at the same time organically interconnected. The person is sound because a vital and aware part of the ever-regenerating, abiding *and* changing, World-whole.

Now we are in a position to better understand the essential role of olfaction.

Scents, Aromas, Bodily **Spirituality

Scents, odors, and aromas open the gate to slippery and powerful bodily **spirituality. This is particularly true when an aroma is inhaled

unexpectedly, its source undetected. An object smelled is not experienced as clearly distinguished from a subject smelling it. Primal
body-selves, we are momentarily possessed by what is smelled. The line
dividing self from its object tends to dissolve, and, at the same time
inevitably, the line dividing self from others *and* the lines dividing present, past, and future.[4] This flowing interfusion constitutes our very being
at a basal level, and though nearly eclipsed from everyday life by practical
routines, prosaic interests, brilliant technologies, and commercial enterprises it does persist, even if today it is not fully empowering.

Blurred as well are lines dividing actuality from possibility, and
human from nonhuman. And with this blurring we tend to feel most
complete, most human, at least sometimes.[5] We are on the verge of
ancient shamanic ecstatic kinship with the more-than-human world.

Eliminating or minimizing smells attenuates involvement in enwombing Earth. As we move into the future, identity cannot be as
steady and strong as it might be.[6] Strangely wrinkled infants come to
mind. Peter van Dresser in "The Modern Retreat from Function"
grasps the basics:

> Beginning by progressively eliminating the need for walking and for
> managing draft animals—two effective modes of contact with the organic
> environment—the automobile has evolved into the "womb with a view,"
> and on wheels: the concrete embodiment of revulsion against a disturb-
> ing underworld of dark fears of germs and physical contact, of exertion,
> effort and dirt . . . The preparation of food, in a parallel fashion, has
> retreated to an esoteric domain of immaculate and hermetically sealed
> machinery, the culmination of which is the inevitable cellophane package
> or tinned container. All the old lusty smells and sensations attendant
> upon the grinding of corn, roasting of coffee, the fermenting of yeast in
> bread or beer, the pressing of apples or grapes, have been banished in
> favor of a hushed operating room asepsis.[7]

We think of tanned, bikini-clad bodies, say, but at an aseptic distance, apart, objectified. Perhaps on stage, or merely depicted in movies or photographs. Bodies involved, but little or no **spiritual
involvement of body-selves living communally and ritualistically and
responsibly with each other. "The triumph of the container," van
Dresser says, is a graphic image of loneliness. Later I suggest that this
mirrors the male body in particular, and his tendency to spectate, to
see self-contained things "out there" to be exploited by his calculations
"in here." He tends to fear something foreign entering and polluting

his pure individuality, destroying the "self-sufficiency" of the container and its contents.[8]

Viscera as Sense Organ

Malidoma Somé is an African shaman so intimately in touch with the body as ecstatically unified sensuous whole that smells are not merely smelled with the nose, but with the whole self, including the hearing self:

> *I want to be what I know I am,*
> *And take the road we always*
> *Forget to take.*
> *Because I heard the smell*
> *Of things forgotten*
> *And my belly was touched.*[9]

But some of our ordinary talk suggests this same synesthesia. We speak aptly of certain scents, odors, or aromas as visceral experiences. The whole body-self is involved in the world through its viscera; our minding does not function merely as an information-processing unit embedded in the skull behind the eyes. Olfaction prompts kinesthetic modes of nonlinear awareness that engage us in the world in the most powerful way. We know our whereabouts because we re-live the sources and places from whence our whereabouts springs, and we feel what is yet unrealized pulling us.

If not badly alienated, we know in our viscera what deserves our care, whether it is immediately before us, or lingers from the past, or comes to meet us out of the future. The biblical metaphor, bowels of mercies, is apt. To know what deserves our care (even if sometimes we don't feel like giving it) is to know ourselves. Attuned to circular power launching itself forward and returning into itself rhythmically, there is no need to frantically "seek ourselves" or "our voice."

As James noted, basic affectional facts are not neatly categorizable on either the subjective or the objective sides.[10] Is the beauty of a rose subjective or objective? It is neutral, or basal, refusing to slip safely into either bin. Where, James asks, is the disgustingness of a bed of maggots writhing and vibrating before us? We cannot say. The bed floats through us in its appalling electric rhythm. Just to name it, *maggots*, is to know it.

Sight alone may be enough to set this kind of experience going, for the senses, after all, function in ensemble. However, analysis of olfaction and kinesthesis best grasps the intimacy of otherness and body-self's strange vulnerabilities and susceptibilities, as well as its sustenances.

Naming the Real
When body-selves name on the ecstatic level of sensuous experience, the names stick. Adam, at creation, is said to have named the animals. The poet names the world, Martin Heidegger observes. The physicist Richard Feynman recalled his father warning him against the seductive power of pretty names, for, like Circe and the Sirens, they lure us into dead ends, lull us from the task of finding out what makes things *go*.

That may be good advice at some point for a physicist, but it risks concealing the power of names to orient within the ongoing world and to energize and inspire us through and through over the long term. For the sensible, nameable world is the abiding matrix without which nobody could begin to inquire into hidden causes of anything. The fitting name seeds and prompts the matrix so that named and namer intertwine, exfoliating and revealing themselves together. Ignore for a moment primal names that curse lives, leaving fixation and addiction, black holes that open any moment.

Today I am to visit my mentor from my graduate student days. He is seventy-five and says he is an old man, and that is true. I have a relaxed lunch and dawdle on the streets of Manhattan, for it is only noon, and he says he wants to see me at three—not two, not two-thirty, as I had suggested. I pass a peeling movie billboard, "Dinamation's Dinosaurs," with a colored plate of a very life-like Tyrannosaurus rex. I get my car out of the parking garage and head slowly out of the city toward his house in the suburbs. On the way I note a looming highway billboard. "Actual Sighs" it is headlined, and shows an infant crawling into a frame, clearly the frame of the video camera held by a man, probably the father.

I am still early, so I stop beside the road several blocks from his house and read a book until three. Then I must hurry because I cannot be late either. This insistence on punctuality is something new. I knock on the door and he shuffles with short little steps to open it. About twenty years ago he had mimicked perfectly this gait of old men, and the memory jostles what I perceive. Once exactly my height, he comes now only to my ear, his wispy hair sticking out in tufts from under a beanie that keeps his head warm. Once, about thirty years ago, I had clapped him on the back in a festive moment. It had been like hitting a tree trunk.

He ushers me in with his characteristic friendly greeting, completely believable as usual, and leans on my shoulder as we walk slowly into his room. Once the dining room, it is now a place from which he seldom moves during daylight hours. On the round table is a hodgepodge of pipes, pouches and plastic bags of tobacco, snacks, magazines, newspapers, empty or half-empty cups and glasses, the crossword puzzle which he does daily, and other objects. A few of his own books are there too, brown and moldy looking, his own copies.

I make a joke. "A look at this table renders Cézanne superfluous." He smiles faintly, tries to locate something in his pocket, fails, looks at me silently, benignly for awhile, and says, "Tell me about yourself."

He really wanted to know. As a graduate student I would sometimes call him just to hear him say, "How are you?" He had always hated to talk on the phone, so would soon say, "I want to see you," and that would terminate the conversation. But he meant what he said. Once on a pay phone time ran out and I struggled to find change. He said something about fishing for coins in pockets being irritating, and that was exactly right—he named it.

In response to his injunction "Tell me about yourself," I began with the billboards in the city. Somehow I could not respond at the invitation's level of sincerity.

"The verisimilitude of our images is so tremendous! At some level we know, of course, that what's imaged is not real, or that, though once real, it's not happening before us now. But don't you think that constant exposure to these brilliant simulations slowly undermines the sense of the difference between real and simulated?"

"You mean we no longer confidently name the real?"

That named it.

We talked about other things, and he got his pipe lit yet again. Whether the tobacco was exactly the kind he smoked thirty years ago did not matter, for everything returned to me as it had been then. My talk ceased. Everything that mattered was held in a radiant circle out of time's way.

Powerful Names

The empowering name is fertile each time it is used, involving the namer fully in the named. It is not mechanically repetitive, addictive— flailing away at a slippery world. It names the real in a kind of sacred bond. As the self names, it is regathered and reanimated again here and now through the named. That need not be anything that exists now in space, need not be real in that sense. It might be a long-departed, revered ancestor whose presence returns supportively, or it might be a former animal companion. It might even be a being that never did exist in space and time as ordinary beings do. But if it is invoked through being named, that is all we need mean by "real." In fact, naming is particularly powerful and ecstatic when it overcomes radical absence in space and time.

In naming we can feel vaguely but powerfully bonded with the World-whole, not confined in our bodies. In the very process of naming things, we connect them, even if they are thought to be disconnected in the world. I name my friends: Joe, John, and Jim; their presence animates me. But I know that Joe is alive along with me in New York, that John is dead and buried in South America, and that Jim is alive in China. But they are together in my naming them. The same with Buppie the dog: he no longer lives, but his presence remains.

Out of the past, presence; out of death, life. And I, the namer, am experiencing myself together with them all—together in a fundamentally real sense, though one that is typically overlooked today. If in naming I were constantly aware of myself reflectively as this being at this moment, I would clearly distinguish my reality here and now from those I name. I would distinguish the status in reality of the three friends and note that Buppie is dead.

But the cardinal fact is that I am not constantly aware of myself reflectively and distinctly. Very often my body is "neutral" or primal, neither a clearly defined subject nor object. Particularly, then, in moments of excitement, mimetic engulfment, empathy, the body in its experience of naming blurs into the named, regardless of the named's status in time and place. As with floating aromas, the body-self loses a clear and vivid sense of its own distinct existence here and now. It follows inevitably that the Here and Now blurs into the There and Then.[11] The presence of the helpful past, coming unbidden, buoys and recreates sacramentally. Loyalty pervades and sustains. As we regain mythic awareness, the hunger to *be* is satisfied.

"That tobacco!" "Pine!" "Jim!" "Ghostly Herd!" "Buppie!" "Sage!" "Donna!" "Bekah!" "Gilbert!" "Daddy!" "Momma!" In these informal rituals of naming, we are rooted again excitedly in ever present origins, sources, at the fecund level of primal body-self.

Even in adversity, or especially then, if I open to the world trustingly, the world tends to flow in around me trustingly, supportively. If I close myself, the rest of the world closes itself, abandoning me—astonishing reciprocity and startling responsibility.

Our very truths are strangely disconnected today, as we will see in the next chapter. We can no longer count on allies, unexpected further-ance, as Emerson put it gnomically, or better, shamanically:

> Character is always known. Thefts never enrich; alms never impoverish; murder will speak out of stone walls. The least admixture of a lie—for example, the taint of vanity, any attempt to make a good impression, a favorable appearance—will instantly vitiate the effect. But speak the truth, and all nature and all spirits help you with unexpected further-ance. Speak the truth, and all things alive or brute are vouchers, and the very roots of the grass underground there do seem to stir and move to bear you witness.[12]

We remind ourselves of James's insight: the world experienced pri-mally and immediately undercuts the distinction between self and other, subject and object. It is just a that, "but ready to become all kinds of whats," and our mode of addressing or questioning the that power-fully influences the what that emerges. Emerson:

> We animate what we can, and we see only what we animate. . . . Thus inevitably does the universe wear our color . . . As I am, so I see.[13]

The Eternal Present

We are vital members or parts because we are vital parts of the Whole, and linear time is only one dimension of the ongoing Whole. As Hock-ing saw, the notion of a beginning of conscious life carries with it a paradoxical reference to something prior to the beginning, and if the impulse which is I is a "racial impulse," there is no reason to ascribe age to it: It is, like energy, always new as on the first day. And no reason to think that where one self begins another stops: different selves overlap.

Linear time is not an absolute, as it was for the seventeenth-century mechanists Newton and Descartes. Time is also cyclical and recupera-tive of sources. We are not locked onto the rails of linear progressions. This is why it was possible for Rozanne Faulkner's "Tina Bopper" to grow into a child again, at least in effect. The addiction therapist saw that the unfinished business of Tina's development was the energy "al-ways new as on the first day" that yearned to consummate itself in her life. And this is why it is possible for some of Harold Searles's schizo-phrenic patients to begin to recover. He was simply there for them over months and years, and they could begin to do what they had never done before: to move through stages of growth from early infancy toward adulthood. He must have been a presence so actual and reassuring for them, that he became, in effect, an archetypal or mythic, as well as actual, mother or father presence. Such presences know nothing of linear time.[14]

The Body's Trunk and the World Tree

The organism functions as a whole. But unlike the head, arms, and hands, the trunk is not even relatively autonomous. A certain natural distance and directedness both connects and distinguishes the hands that touch from what is touched, the ears that "cock" from what is heard, the eyes that focus from what is seen. Much less distance divides the experiencing trunk from what is experienced by it (although there is a certain directedness of the trunk toward what is experienced by it; for example, our whole trunk resonates to vibrations, whether or not we can locate their source). So our visceral or, most generally, our trunk activities are much less able to distinguish their own reality here and now from what is experienced by them, much less able, then, to get any perspective on themselves (less able than the organs in the head that can easily be directed voluntarily toward their present objects by turning the head alone).

And so it follows that the Here and Now in visceral experience tends to blur deeply into the There and Then. Having always circulated within Mother Nature, the matrix of the world, the sound and vital trunk holds our deepest on-tap spontaneities, our gestatings and birthings in the moving darkness. Particularly, coiled intestines intertwine with energic beings on the magical level of experience, as mythic presences—totem animals, fecund goddesses, gods, sundry allies—cycle out of the past of the race and leave their living residue in us.[15] Flowing in from the past, they beckon toward the future—a perpetual cycling and buoying flow. The self is moved and consents to be moved. If we are viscerally assured of reliable coherence and endurance here, there is no need for addictive pastimes and fillers.

Visceral and trunk responses have both a great strength and a great drawback. Their strength is obvious. Not needing to be directed toward what is over there at the moment in space, the trunk is centrally involved in fusing a life together through time. Moreover, the trunk needn't respond to anything that was, is, or will ever be simply "over there." That is, it needn't respond to merely localized reality. It responds moodily to whole fields of interdependent things, whole vibratory and rhythmic situations, even the World-whole. It responds to what may never be visible or definable.

So the whole trunk can be the ever-replenishing center of energy and elan—of our being as creatures of the Whole—that carries right through black holes in time. With this momentum we disregard the whispered temptations of immediate gratifications. Jean Gebser writes of "the numinous role of our [visceral] organs in the awakening of consciousness."[16] I am supplementing this characterization.

But today of course we tend to live without orienting and empowering rituals, and the main drawback of the trunk is that it may close in upon itself and fall into darkened ruts of addictive repetitions. Constant and contrastless, so close to us that they *are* us in an equivocal way, addictive activities are often difficult to detect, let alone to root out. As if body-self were repeating, "I am doing something, I am doing something, I exist, I exist"—the pathetic tediousness of addiction.

The trunk shows greatest derangement in addictive eating disorders. Anorexia and bulimia are terrible hungers, for they attempt the truly impossible: to satisfy our deepest need, which is to *be* fully, through finding a substitute for ecstatic action, differentiation, and fusion in the world. But there is no substitute. Anorexia and bulimia are bad magic—attempts to reduce body-self to a mere body that is stuffed, followed by attempts to deny what one has done by disgorging the food and drink (bulimia).[17] Or, one fails to face the ecstatically impoverished body-self and forces the belief that sheer mental control of eating is enough to maintain a life (anorexia).[18] Body-self achieves no overarching integration and belonging, and black holes in time open up. Refusing to swallow, or refusing to stop swallowing, anorexics and bulimics are swallowed by the world before their time. Agency, freedom, self is lost.

In these disorders the body with its viscera and trunk is no longer connected in the World Tree (Yggdrasil, as Emerson retrieved it from Norse myth), the community of all procreant life that sends out branches into the heavens only proportionally to how deeply it sinks roots in the darkness of the Earth. Those afflicted with eating disorders no longer trust the interdependent world, no longer trust the regenerative rhythms of the Earth-and-Sky that formed body-self's appetites and understandings.

As the child plays with its fist in the field of vision, so adults can explore and play. We might learn to trust again, to move out into the unknown, just because it's there, to find the self-correcting momentum of that ecstasy. We turn now to art-making: activity coeval with the human race that taps "energy new as on the first day." It is a good magic in which we "do the impossible"—make the other our own, as it makes us its own. Caught up in the compatible and fertile other, we are "me and mine" through and through.

Walt Whitman in "Song of Myself":

> Urge, and urge, and urge,
> Always the procreant urge of the world.

Notes

1. *The Spirit of Modern Philosophy*, New York, 1983 [1892], 7.
2. By thinking of *now* we are typically able, of course, to mean what is happening simultaneously at a great distance from ourselves. For example, "It must be 7 a.m. in Moscow, so Martha is probably getting up now." I am only saying that this mode of consciousness, so essential to our technologically outfitted lives, does not always dominate our experience.
3. See Frank P. Jones, *The Alexander Technique*.
4. The structure of the portion of the human brain involved in smelling is what one would expect on the basis of the immediate experience of blurred distinctions between present and past. The olfactory bulb fed by the nose leads directly into the old brain, the hippocampus and thalamus, and the stimulus may never get to the cortex, the seat of most articulable distinctions (I am indebted to conversations with Gilbert B. Wilshire, M.D.).
5. In his valuable *Coming to Our Senses*, Morris Berman discusses some of these themes. But he thinks of bonding with others as "syncretism," a "confusion" of self and other. But not all bonding is confusion. We *are* field beings caught up in corporate individuals (and not just in "force fields"). We *are* interfused beings, and *also* individuals.
6. A reader of a draft of this book has pointed out that authoritarian persons often try to disembody others by denigrating the smells associated with them: "I'm through and through Ukranian by ethnic origin. Garlic is essential to this way of life, yet my father hides it in plastic in the basement, and I leave only the uncleaned garlic press in the sink after dishes."
7. "The Modern Retreat from Function."
8. See Northrup, *Women's Wisdom*, and Schaef, *When Society Becomes an Addict*.
9. Rendered from the African Dagara language by Somé, who goes on, "In Dagara you 'hear' smell, you don't smell something. That which is picked up from the olfactory sense is sound being heard in that way." *Of Water and Spirit*, 296 and fn. This indicates the rhythmic, vibratory, and visceral nature of life.
10. "The Place of Affectional Facts in a World of Pure Experience," in *Essays in Radical Empiricism*, and excerpted in my anthology.
11. Emphatically: this blurring is not necessarily a fault, a blemish, a confusion.

It can be the deepest bonding of self and world, a necessary condition for the self's substantiality and vitality. Morris Berman, after Michael Balint, describes our emergent awareness of the other, and of ourselves as seen by the other, as the "basic fault": We parade before the other and are divided from our visceral sense of ourselves; we are split (*Coming to Our Senses*, 36). But this is not inevitable. True, we become rudely aware that an interpretation *might* be put upon us that is antagonistic to what we immediately feel about ourselves. But without primal experiences of bonding positively to others, of being understood by them, the fact that we can be misunderstood by them—that they can attempt to erase us—would be unendurable.

12. "An Address Delivered Before the Senior Class in Divinity College, Cambridge," in Ziff, 109–110.

13. "Experience," 289, 307.

14. By *archetype* or *archetypal* I mean archaic patterns of behaving and relating to the world that can be tapped in our present experiencing and which need not be limited to a particular culture. C. G. Jung adapted the term from the Greek (*arche*, origin or source; *type*, what is produced by a blow, as the imprint of a coin; so the word suggests "primal imprinter"). Jung anticipated modern ethologists' notions of stimuli as "releasers" of "imprinted" patterns of behavior. James Hillman writes, "Let us then imagine archetypes as the deepest patterns of psychic functioning, the roots of the soul governing the perspectives we have of ourselves and the world" (*Revisioning Psychology*, xiii, cited in LaChapelle, *Sacred Land*, 372). Jung's intent, I believe, was to understand the archetypes "as the animal and earth itself within us" (LaChapelle, 74). Jung wrote, "The symbols of the self arise in the depths of the body and they express its materiality every bit as much as the structure of the perceiving consciousness" (*Essays on a Science of Mythology*, 92). However, he could not free himself from dualistic formulations. In their most crippling form: consciousness or *thought* is associated with "the Masculine," while subconsciousness with "the Feminine." (See Annis Pratt, *Dancing with Goddesses*, 5–6, 75). Where does this leave women who *think*? Are they merely topics to be excogitated in men's thinking? Indebted to James and Pratt, I prefer to speak of archetypes "neutrally": Archaic, recurring, evolving patterns of human-plant-animal-earth-and-sky interfusings. They are the analogue—usually dimmed and attenuated—of Aborigines' Dreaming. Paul Shepard writes wisely, "Learning . . . means a highly timed openness in which the attention of the child is predirected by an intrinsic schedule, a hunger to fill archetypal forms with specific meaning. Neotony is the biological commitment to that learning program, building identity and meaning in the oscillation between autonomy and unity, separateness and relatedness. . . . [Learning] is a pulse, presenting the mind with wider wholes, from womb to mother and body, to earth, to cosmos," *Nature and Madness*, 110. I maintain in chapter 11 that whoever plays the motherly role is caught up in the archetype Mother.

15. Jean Gebser, *The Ever Present Origin*, indexed under Intertwining.

16. 145.

17. Becky Thompson writes of a woman (*A Hunger So Wide and So Deep*, 78),

"When she was a young child (she) binged because it made her 'real self disappear,' so that no one could find her. Many survivors of sexual abuse describe a similar splitting process. Whenever she was physically or verbally attacked . . . she would disappear. . . . She described this safe place inside her body as filled with huge balloons made of globules of fat in cellophane. When she ate she felt . . . she could hide in this room." She treats the body as if it were a fortress to hide in. But if the body is the self, this uses the body as a *means* (even if to save oneself), not as an unconditionally valuable end-in-itself, worthy of reverence. This must cause guilt. Vomiting after gorging is probably a way of atoning for the guilt.

18. Susan Bordo writes (*Unbearable Weight*, 178): "Anorexia *begins in* . . . what . . . is conventional feminine practice," but in the course of dieting "the young woman discovers what it feels like to crave and want and need and yet, through the exercise of her own will, to triumph over that need . . . a new realm of meanings is discovered, a range of values and possibilities that Western culture has traditionally coded as 'male' and rarely made available to women: an ethic and aesthetic of self-mastery . . . expertise and power . . . through superior will and control." The Male archetype of detached and focused mental power is not limited to male bodies, but sweeps through a culture, disfiguring female body-selves.

Chapter Ten

Art and Truth

Art thus represents the culminating event of Nature as well as the climax of experience.

—JOHN DEWEY, *EXPERIENCE AND NATURE*

A painter told me that nobody could draw a tree without in some sort becoming a tree; or draw a child by studying the outlines of its form merely,—but by watching for a time his motions and plays, the painter enters into his nature and can then draw him at will in every attitude.

—R. W. EMERSON, ''HISTORY''

One of the great insights of hunter societies is that words and artifice of specific place and place-beings (animal and plant) constitute humanity's primary instruments of self-location . . . for mankind is fundamentally an echo-locator, like our distant relatives the porpoise and the bat . . . Only by learning . . . *true* words and *true* artifice about these things can one hope to become . . . a genuine person. . . . To be mendacious about other-than-human persons springs back upon us to make us mendacious about ourselves.

—CALVIN LUTHER MARTIN, *IN THE SPIRIT OF THE EARTH*

Many years ago, my brother and I began an ascent of a high pass on the eastern face of the Sierras. The climb was to be about 6000 vertical feet in two days, beginning at road's end at 6,400 and cresting at the pass at 12,300. In early

morning light, with full packs, we started from clumps of dusty bluish-green Live Oaks near the rushing stream.

Dormant muscles sprang into play, exhilarated; the body hummed. Redolence of oaks, then pine, then the smell of the newly risen sun on rocks and earth, even the warmed dust buoyed our steps.

Thinness of oxygen was already apparent, but we kept pounding upward toward a narrow canyon of precipitous rock walls. Purple, tan, reddish the rock was, with veins forming whitish stripes against dark stone, like the rump of a zebra. To look back was to see the Live Oaks achieving miniature status, and the desert valley below opening out, the very dry high mountains beyond rising and falling into the state of Nevada.

We had climbed about 1,500 feet. The sun now shone hotly. Stopping for several breathers, we kept our packs on—leaning them and ourselves against boulders and rock walls, unwilling to let the tension go out, and to have to reassume the packs and regain momentum and rhythm.

As we slogged upward, the world continued to open up behind us and the vista in front promised the still-distant pass. Pines were gnarled dwarfs, the sign of the approaching tree line that would appear at about 10,000 feet. I stopped for an instant beside a gooseberry bush that extended its thorny branches close to the trail. Carefully picking a half-dozen ripe berries, I noticed that my fishing rod, protruding from the pack, cast a shadow on the rocks.

Tasting the berries, a wave of nausea passed through me—a spell of altitude sickness. The berries trickled and fell. I stopped, legs shaking, but then went on anyway. After about ten minutes, however, fearing total breakdown, I sat down in a heap with the pack still on.

Sucking in air, holding myself like thinnest porcelain, I requested my stomach not to rise. My brother looked at me out of the corner of his eye. For a long time it could have gone either way. But luck was with me, and in about forty-five minutes we continued on. After several hours we finally stumbled to a camp at around 10,500 feet—so our topographical map showed. Too exhausted to take more than a cup of soup, I slept fitfully, as consciousness, like a dry light, sporadically faded. The bright stars stepped their way across the sky.

Morning came cold, the granite walls steel gray. Quickly we breakfasted on nuts and raisins, not wanting to dirty aluminum pans, eager to cover ground before the sun hit us. Up and up we went through a world of rock. White granite faces of peaks stood out against the cobalt-blue sky. We stopped frequently after we reached 11,000 feet. Still the trail switchbacked up ahead.

The goal of the pass caught us up in itself and possessed us, up, up. Around the agony billowed tremendous excitement and contentment, as if we were graced by some deity. Life was a coherent surge that retained its past and moved into the future. There were still peaks above us, but more and more the world opened, the circle of the horizon expanding over thousands of square miles. The line of the pass—unless it were a false line that concealed still other lines behind it— now stood just above us, perhaps no more than three hundred yards ahead.

The trail eased its ascent and we hastened to the crest. Peaks on the other side showed their heads, and then we were there: a whole vista of new peaks and valleys and clouds, another sector of the world. The complete circle of the horizon could not be seen. But enough of it could be to sense the everything else, *the whole on every hand beyond the horizon. It was the universe itself, arrayed, and my brother and I vital parts of it.*

After a leisurely lunch at the top, coats and sweaters on, we coasted down the other side to our goal—Baxter Lake flashing in the sun, a study in blues and silvers, carved into a granite shelf by glaciers many millennia ago. Snow-streaked granite peaks rose on three sides, and scattered clumps of aromatic pines no taller than ourselves grew bent and carved by the westerly wind. Thunderheads boiled over the peaks, and after dropping the pack, I found a round boulder, a perch on the world to sit on, quietly elated.

Like Emerson, I felt on familiar terms with insentient things. They were not mere objects in the "outer" world. John Muir could speak of the onlooking mountains. I too—I was addressed by them, as if they were faces and voices. As I fed my elated regard into them, they fed back theirs—elated interfusion. Not just "its," but "thous" and "yous" we were, addressing each other. This intimate fitness of things—this humming and whispering of belonging—was beauty itself.

Was it possible to feel happier or more fulfilled? The question itself was probably impossible in those moments.

A few years later in the mountains of Southern California, the gentler San Gabriels: I gained a ridge and found a painter with easel and oils painting the crest of that range. I was fascinated but could barely contain laughter. I had achieved that ridge, I was taking all that in, that vast scene and mood. Why try to reduce it to a square foot of canvas daubed with pigments, lumps of colored mud smeared into shapes? It was redundant, or worse than that, an insult somehow. As the painter worked, my fascination alternated with contempt. Are some people never happy? Then came a glimmering of what he was doing. But I could not grasp it.

Why had he been doing that? What might I have been missing? What might I still be?

Over the years I came to see that by placing blank canvas, recalcitrant blobs of pigment, brushes between himself and the scene, he was becoming intensely aware of the feeling, moving, coiling and uncoiling body-self itself—primal, serpentine, more labile and powerful than body as object or self-conscious, self-managing subject. His feet planting him in the Earth, hand and brush darted out to deposit its load of pigment. Moving and intent head, molding and directing hands, viscera with its coiled intestines alive and alert—all this spontaneous activity was becoming aware of itself as the painter worked-up his materials.

I saw that the infusing, glowing scenes around me, which I thought I had taken in completely, partially eclipsed an element absolutely essential for generating my experience of them—my own body. Even here the body was realizing only part of its capacities as body-self. I was taking in the scene and making sense of it, but not on the ecstatic level at which the painter worked.

He was making sense of his making sense—grasping bodily how meaning is made by this very body in concert with its surround. And why not suppose that these ways of sensuous meaning making are used in everyday experience, but not thematized and prized for their own sake? And that art draws them out of the margins of consciousness ecstatically and into the focus of new life?

Art Breaks Open and Cultivates the Ground of Meaning
Sitting on the rock in the Sierras and achieving the ridge in the San Gabriels—each was *an* experience, a unity, a successful regenerative cycle. Suspense over whether I would make it gave way to elated consummation which, in turn, deferred to quiet satisfaction and consummation savored. An implicit rite of passage occurred—"I can do it!"—and if it had happened at a crucial turning point of my life, as thematized by the culture, it might have been explicit.

Without such dramatic unities (even bending over to sniff a flower is one such, or taking in a scene), there would be nothing for fine art to develop and illuminate. Human identity through time is an achievement: we don't just have our identity as a number has its. But such fruitions of experience are not yet fine art.

Successful fine art reveals the conditions of such consummatory experiences, what makes them possible—our artist-artisan bodies world-involved. As we take from the surround and feed back into it, molding materials, we create some new thing. Fed back is awareness of what we fed in, prompting what we next do: A self-correcting and self-energizing circuit that is never fully predictable.

Others can react to the art, identifying with it, experiencing themselves and the rest of the world through it. And their reactions, actual or anticipated, are fed back into the self, vastly augmenting the feedback, reflecting and confirming what the artist fed in, and how it was. We are body-selves creating meaning in improvisatory concert with the world around us; ourselves-augmented-by-the-world replenishes us. With will integrated with desire, judgment with need, impulse, and intuition, there is no opening in the self for addictive fillers, no equivocal "me and not me" experience. The painter gives birth to the painting, but only as the world gives birth to himself.

Differently put, the painter gives the scene and its horizon back to itself through that part of it that is his or her own organism. The world births the painter, and at the same time enriches and recreates itself. This is a vivid corollary of the general principle of value: individuation of self through being a vital contributing member of the self-regenerating World-whole.

In the creation of a work of art the conditions of an experience flow into awareness—the birthing matrix unfolds. And we realize that this is not something subjective, cooked up "in minds," which is then "expressed." It is what Nature imparts to us about us and our evolving place within her growing totality. It is a pigment-earth-sky-water-brush-body-speech that has changed but little over many millennia (note the cave paintings of animals made thirty thousand years ago at Pont d'Arc, France, and ancient sites in southern Africa). In discovering what the world makes of us, discovering where we stand, we echolocate ourselves, find who we are and what we are becoming. Truth on this level is not the desiccated correctness of prosaic ideas and speech. It is truth as body-self true to itself, true to its possibilities for integration, power, and freedom. It is body-self living at the farthest possible remove from addictive cravings.

Art opens us up to the unpredictably moving and gliding primal body circulating within the rest of the world. If it fails to do so, artistic activity is mere diversion, not the ever-growing formation of meaning and of ourselves, not an essential root of culture. We would merely repeat what we already know.

Artists take us where we've never been, or never been with full awareness. As I discovered on Mount Goddard, possibility and actuality merge on the fearsome ecstatic level of primal body and world. The possibility of falling is a strange actuality. The line between human and animal experience blurs. Presences and meanings that haunt are not necessarily either here or there, now or then; mine or yours, human or animal; not necessarily potential or actual. Artists animate possibility at the fecund mythic level before it is thematized prosaically as possible, or, possible, not actual.

Artists are kin to shamans, creating in states of regenerative trance or semitrance. Control lies outside the calculating, fragmentary ego-self. Adepts, they allow the larger world to move in and emerge through them oracularly. In art we find ourselves through the kindred that resonate to us—unpredictably gifting—vegetable, mineral, wild or domestic animal. Body-self and the rest of Nature circulate through each other in a womblike way, but now the circulation is artistically articulated and responsible. It is some evidence for maturity.

Might this circulation begin to permeate one's daily life in the form of ingrained regenerative ritual? Might we achieve the vivid awareness that every moment may be our last and be eager to live?

Too often today living is a series of disconnected trance-like involvements.[1] Art is the countertrance that jogs us from rutted performances and opens a larger and deeper reach of who we were, are, and might be.

Art and Regenerative Experience

Art and religion grow from a common root that is paved over by secular-commercial culture. They are mythic forms of life, ways of participating in ever-present regenerative processes of Nature and disclosing their unconditional value. Take Andy Warhol's enlargements of Campbell's soup cans, or iterated, identical depictions of them. These may seem merely perverse attempts to capture attention, crotchets, or merely addictions. But if we keep looking we might see something else. This fanciful detachment from routine reality can reanimate the presence within us of regenerative things, in this case a "simple" nutrient, soup. And when he repeats the same picture of Marilyn Monroe? Sultry half-closed eyes, opened mouth, her steamy presence—all this easily becomes mere sex in talk about the art and its market value.

But on another level of meaning, mute probably, don't these uncounted iterations suggest inexhaustible reproductive and regenerative power? However unsure intellectually we may be, don't we fleetingly intuit that Marilyn Monroe plays for us a Goddess role?

In the art of aboriginal hunter-gatherers there is no such hesitancy and division of mind. Chris Knight comments on a bark painting of the Rainbow Serpent—for Aborigines the ever present Dreamtime source of life. In a sequence displayed simultaneously forever, the serpent smells two synchronously menstruating women, pokes its head into their ceremonial hut, engorges them, slithers away, the women still visible inside. (Objectifying, reductionistic, North Atlantic critics call this "X-ray style.")

> What creature on earth . . . could connote cyclicity more appropriately than this flowing being, coiling itelf up in spirals, undulating its way across water or land, sloughing its skins. . . . Appropriately for a menstrual metaphor, most snakes have a quite extraordinary sense of smell; it is after smelling the two synchronously bleeding Wawilak Sisters' blood that the great copper-python Yarlunggur . . . incorporates these dancing, synchronized women into its body. . . . Water was certainly central to the rainbow snake, but . . . this was . . . sacred water . . . womb fluid—the menstrual flow mingling in myth and imagination with the surrounding streams and waterholes on which life . . . depended, swallowing up men and women in its synchronizing, rhythmic power . . . Almost anything could be a part of rainbowness/snakiness. Just as the arch of the rainbow mediates between earth and sky, dry season and wet, sunshine and rain . . . so ceremonial life across Australia . . . had carried humans in an orderly way from one season . . . to its opposite, from the "raw" phase of each ritual cycle to the "cooked," from blood to fire, kinship connectedness to . . . marital life. . . . Ritual . . . activate(ed) all living being as vibrant participants within "the Snake."[2]

Human existence is never simple. Aboriginal men tried to expropriate female prerogatives through simulation, strange mimetic engulfment in a male corporate body: the penis is slit, tied against the abdomen, its oozing blood simulating the female genitalia menstruating. The story of the consuming serpent was used to frighten women.

But the universal core of the story in the bark painting is salvageable: body-selves and the rest of Nature live reciprocally. We grow up within Nature, nourishing it and being nourished by it. To know the force and sinuous course of life, we must feel it in our bodies, and in the possibilities charging the atmosphere. Until finally we have given and gotten all we can, at least consciously: individual centers are consumed, recycled.

We live in a fractured time. Dewey put it well: "The world seems mad in preoccupation with what is specific, particular, disconnected."[3] We get no clear reading of how Marilyn Monroe, childless, probably a drug addict, perhaps a suicide, could lead us back regeneratively into Mother Earth. But are we totally cut-off—caught in the specific, particular, disconnected—mad, as Dewey said?

White man have no dreaming
Him go 'nother way.[4]

I am not optimistic, but not totally pessimistic either.[5]

Art and the Healing Organization of Space and Time
Dewey writes with penetrating concision,

> Form, as it is present in the fine arts, is the art of making clear what is
> involved in the organization of space and time prefigured in every course
> of a developing life-experience.[6]

Art unpacks our prefiguring of space and time to reveal and charge
our enculturated **spiritual bodies. Only through habits in which we
conserve meaning do we in-habit the world, as Dewey puts it, and only
when able to create more meaning do we in-habit it vitally.[7] Otherwise
we are mere objects rattling around responding to stimuli and craving
momentary gratifications. Successful art inserts us living organic grafts
into the ground, trunk, and branches of the world. *Contact! Contact!*

*We were in a rented car in the interior of Australia. Traffic goes on the left side
of the road here. My wife was driving, for I feared my attention might wander,
and if jolted by something, I might swerve into oncoming traffic.*

*Between stretches of monotony were times of intense coherence and fascina-
tion. Details on the distant horizon were a bit softer than features close at hand,
yet etched with uncanny clarity. In touching the horizon moment after moment
with my eyes I was also touching something ancient and intimate in myself.
Space and time were prefiguring themselves through me. They were the encompassing ground and ever-regenerating root of myself as body. What made an
experience possible was itself an experience.*

*That is, what I always already was as the world's creature arose alive and
quivering in consciousness, seeming both inevitable and spontaneous. I felt on
the verge of art. The best I could do was to hum, then whistle. I broke into
laughter. My wife's quizzical look encouraged me to think that I could, with
effort, explain what I was laughing about, but I was too happy to try.*

To be perpetually set to experience the world spatially and temporally
and ourselves evolving seamlessly through it are deepest habits. But
even they wobble in situations when the sacred alltogetherness of things
is lost. Then we may pitch into a black hole in time, when all we can
imagine intensely is what will immediately satisfy our needs—cure our
depression, remorse, desperation, whatever ecstasy loss afflicts us *now*.
We collapse into addictions.

Regenerative Conceptions of Truth

Art, ritual, and intrinsic value flower from the same stem. In making art we celebrate our creative powers because the world creates sectors of itself through us. When we ritualistically repeat some event formative of our group in the cosmos, art becomes explicit religious ritual. We experience the intrinsically valuable excitement of growth rooted in source: individuation most vital because most enhancing of the Whole, unconditionally valuable, mythic. We are true to ourselves.

Most thinkers in North Atlantic culture today have lost this most basic sense of truth. Having bought into the technological world of the detached, calculating ego-self, not considering that art can be true, they express contempt for myth. Typically they assume that truth can only be a property of prosaic declarative statements or propositions.

This development was foreshadowed by Aristotle in Book Epsilon of *Metaphysics*, where he says that truth is a property of statements or judgments "in mind" or "in language." Statements have this property when they assign predicates to subjects in a way that corresponds to the way attributes are possessed by the things "out there" that are judged about. We judge, say, that the apples are yellow and they indeed are yellow. Two domains, mind (or language) and the world of things, match up, supposedly.

But we can use Aristotle as a stepping stone into the more distant past. Remnants of a much older way of conceiving truth appear in Book Delta. There we find that a *thing* is considered false if it habitually appears to be what it is not. So a thing can be true if it habitually appears to be what it is. Truth is truthful showing or revealing of things, perhaps mute.

Integral to the development of living things are the appearances that truly show and disclose them. Maturation requires communication. Understood in the most general way, then, things (not just statements or judgments) can be true when they are true to their developmental capacities in the fertile world, true to themselves.

This line of reasoning suggests that fine arts are true when they facilitate persons' and things' developing and showing themselves in their tangible reality, their sensuous meaningfulness and valuableness. It is easiest to note truthful showing in the case of representational or "realistic" art forms. A cardinal value of a great portrait is its revelation of the sitter's character, his or her impact on the painter and ourselves. Fine art has this normative notion built into it. By establishing the dignity of showing, fine arts prevent a cut being made between reality and appearance, with the latter typically denigrated today as *mere* appearance or subjective experience. They prevent the isolation of mind (or language) from the pulsing, ecstatic world of things, prevent loneliness or vertigo or grim productiveness (one more conquest, one more widget, one more fact). When things show themselves as what they are, in their truth, surfaces and depths feed each other within experience. Beings live in communion. When the mountain pass shows itself truly, the Earth flows back to itself through us, feeding us. Things allowed to emerge in experience and be what they are compose for us a kind of sacrament. This is the deepest freedom, for it washes out the deepest blockage to development: dread of disclosure.

Now, when fine arts enhance things' ability to show themselves truthfully, they participate in their truth, and are true themselves vicariously. We know this happens when various forms of showing (including relevant prosaic descriptions if available) converge and confirm each other: they work us into more fruitful and thankful ways of interacting with things.

Symbolic and Mythic Art

Symbolic and mythic art realize and exhibit profoundly regenerative human involvements in the world. There is no single, demarcatable object out there we can compare them to, as we can compare a sitter to the portrait. To get at the truth of this art and design, and the reality it bodies forth, we must resurrect an even more ancient conception of truth as *aletheia* (*Lethe*, the river of forgetfulness for the Greeks—*a-lethe*, unforgetfulness). Whatever keeps meanings *alive in memory* is true in this sense. And art can do this, even when we secular people today would probably say it is false in the prosaic correspondence sense. In art we remember bodily the world in which we were formed. Our possibilities of being true to ourselves are fed.

Even when ancient senses of truth do not cohere with the correspondence truth of contemporary science (and a perversely limited common sense), they compose a context for locating and delimiting this truth. It can no longer pass as the whole story. It contributes to disclosure of ourselves, but with its exclusivist pretensions stripped away.

Though lacking important scientific and prosaic truths, indigenous peoples could hold together prosaic and art-formed truths in one throbbing, mythic story of life. Most in North Atlantic civilization today tragically cannot. Artistically and shamanically evokable realities of appearance and presence are paved over or dislocated.

But by retrieving remnants of ancient intuitions of truth through molding materials, we reclaim to some extent an ecstatic belonging in the Earth. We echolocate and realize ourselves when we allow things to realize themselves, to respond fully to us. Simultaneously, we are shown in our truth: true to ourselves when we let things be true to themselves. When we celebrate this capacity in art we verge on the reritualizing and remything of our lives.

Where would we be without this openness in the world, where molding, showing, and development happen—where a ceramic pot or sculpture or music emerges, completely without words? We would not be able to even imagine or fancy that words or mental states match up with physical ones "out there."

Modernism and Scientism: Uprooted Truth

Most theories of truth, particularly since the seventeenth century, are abstract and alienated. They stem from the exclusivistic mind-set of objectification and experimentation—ecstatic in its own way—that forces Nature to answer our questions. And for this monopolistic advance the price paid is the desecration and exploitation of the Earth and our own bodies.

These Western theories focus exclusively on the idea that truth is correspondence between inner mind (or writable language) and outer world. In the absence of archaic modes of truth in which self realizes itself ecstatically—realizes itself by communing with itself in the very act of communing with the rest of the world—more and more prosaic correspondence truths must be amassed. This is a black hole of dependency. We can never get enough of what we don't really want. We are stuffed with more facts than we can make our own, and still never fill the emptiness.

Since the seventeenth century, science and secularism have tried to correct for religious-artistic "obsession" with a world beyond this one. But they overcorrected. Humans are active and reactive bodily beings possessed in ways by things or beings, not isolated theatres of consciousness, populated by detached images and prosaic statements, running along linear tracks of space and time. All such accounts subtly externalize us, disentraining us from primal experiencing in which we make the world-experienced our own—and, womblike, it makes us its own.

Let us pause a few moments to consider philosopher-scientists' correspondence truth. If we were to draw a picture on a blackboard of the conscious mind's ideas and images (these being inside "the mind" and the mind inside the head), and then draw outside this head items in the world, it would be all too easy to compare them. We could see whether ideas and images inside correspond to items outside. If they do, we have truth; if not, falsity. The blackboard enables us to compare directly the inside to the outside.

But if all thoughts are inside our heads, as the diagram represents, then the thought of something outside our head is really only inside our head. If we take the diagram seriously, we see we cannot have what it claims we have: a connecting medium that enables us to compare what's inside to what's outside. But in actual life we can and do check our thoughts to see if they disclose the world. So there must be something wrong with the diagram.

In fact there is nothing in the actual world that corresponds to the blackboard itself! There is no such simple and static connecting medium out there that could allow the direct comparison of mental and physical entities (or linguistic and nonlinguistic ones). Our nostrils filled with chalk dust, we tend to miss this.[8]

The diagram that pictures truth as a correspondence at an instant between mind and world has been drawn on a blackboard that itself corresponds to nothing in the world. The apparently sober idea that truth is correspondence is really addictive assertion of an occult relationship.[9]

The Seductiveness of Correspondence Theories of Truth

The apparently innocent belief that truth is exclusively "timelessly true propositions" is scientism and viciously alienating. It appears to appeal to common sense, to wit: "But isn't the proposition 'Buppie is in the garage' true whether anyone knows anything about this or not? What is this experienceability and meaning business?"

However, it is a perverse appeal to commonsense, for it undermines the commonsense to which it appeals. Note: to be true or false propositions must first be meaningful. And meaning cannot be divorced from our sense of the experienceability of things the propositions are about, what they would be experienced as, if they were to be experienced. Overlooking the absolute priority of meaning making over truth, correspondence theorists—sunk in a weird materialism—simply assume that statements or propositions are self-subsistent entities of some kind with an exotic property, truth, regardless whether anything is known about them, even their meaning.

But without a living situation of memory and anticipation in which meaning is made by bodily beings—a situation that prompts the stated proposition or poses a question to which that proposition is a response—there would be no sense in making it. The statement or proposition has the meaning or sense it has because of the situations in the world—verbalizable or not—in which it makes sense to make it. The sense of these situations is the ancient and evolving matrix of common life and common sense. When the statement or proposition with its "property" of truth presumes independence of the matrix, it undermines common sense.

The totalizing pretensions of sweeping correspondence theories of truth do not merely occasion mind-numbing exercises in a few classrooms. They contribute fundamentally to the constrictedness and uprootedness of North Atlantic culture. It is now widely assumed that every state of reality, every situation, is a state of affairs about which some proposition is true; the search for a theory of truth has ended. "What other sort of truth could possibly be needed!?" Missing is any articulable sense that anything is missing.

Something *is* missing from the "correspondence" account, however. It is what can never be fully objectified and articulated in moment to moment experiencing, for it is this experiencing itself—the developing core of our lives. Missing is the ability to acknowledge and value our own immediate ecstatic involvements in the world.

When truth is believed to be only of or about some state of affairs experienced or experienceable out there, our own experiencing through time, our immediacy and inwardness, becomes an inner out there. Our core shrinks, separates within us, and dries out.

Losing contact with immediate experiencing is the proto-addictive state. Experiencing is mine—indeed, me—but it is also not mine and not me because not ecstatically appropriated and fully allowed. The most primal of needs—to be a whole, vital self—is not met, so substitute gratifications are addictively excited.

For intellectuals these gratifications are typically more and more correspondence truths, and since such may be valuable, a rationalization for refusing full experiential involvement arises that is practically impenetrable. Not rarely one sees intellectuals manifestly at odds with their own body-selves—constricted breathing, tense body—as the self rebels at its own tyranny of itself. Merely counting these people does not indicate their power, for they largely compose the pundits and opinion makers. And they incite a wild attempt in the mass media to counterbalance the pundits: flashing images, sensationalism, gratuitous shocks, instinct divorced from context, mere brute willfulness.[10]

Meaning

Prosaic, factual, so-called correspondence truth certainly does exist as one level of truth, and it depends on prosaic meaning: anticipated consequences of our habitual and routinized postures and actions. We tag things, distinguishing them for purposes of classification and control—meaning becomes conventionalized, a sort of tool. "The dog is in the garage" prepares us to cope with a typical local situation in the world.

But prosaic meaning presupposes an archaic matrix of meaning that it ever inclines to forget: stores of archetypal presences—tendencies to make meaning encoded in our bodies at the mythic level—waiting to be touched into newly developing life. This is the recollecting cultural and artistic ground, which includes science at the all-important level of hypothesis formation. As Niels Bohr has it, "Science progresses by image, parable, and poetry"[11]

In sum, the prosaic statement "I feel excited kinship with the dog," true in the correspondence sense, cannot be adequate for truth. For it objectifies the terms in the relationship, insinuating a distance from the dog that tends to constrict and dampen the excited kinship it tries to convey. This kinship must *show* itself in its archaic truth, disclose our developmental nisus and possibilities by engaging them. Correspondence truth cannot express the full reality. Detached, bereft of this ecstatic showing, we settle, lily-white, into the spectator's chair. Missing is any clear sense that anything is missing—only vague hunger and restlessness remain.

Buppie the dog was and is both an actual and a mythical creature. This is not an assertion that needs to wait on argument or evidence to be confirmed. Its truth is in the meaning of the life's experience it conveys. His body, translated through mine, generates meaning that is absolutely self-sufficient. His value was, is, unconditional, mythic.[12] That is to say, the dog was also Dog.

Meaning and truth are identical in art. Successful art is the arresting showing of the sensuous meaning and experienceable value of things. Either the art circulates lovingly through our neural and muscular proclivities built up over many millennia of adaptation within Nature—impulses to play, explore, mold materials—or it does not, and that's the end of the matter. If we really look we will find these impulses: half-hidden archetypes, perceptions, urges to spontaneous activities.

Erotic Meaning and Truth

Art springs from ancient funds of meaning making pregnant with antic-ipations of what will happen. But we need not go anywhere and look to see if a prediction is fulfilled—to a garage, say. We know what we mean. In successful art we are already in touch with what we want to intensify and clarify, show or reveal—although we can't predict exactly what will emerge. Already in the openness the numinous alltogetherness of things animates us lovingly. We are eager to live.

Emerson says that things in their sensuous presence "pre-exist, or super-exist, in pre-cantations that sail like odors in the air." This can only mean that body-encoded anticipations of things fuse with them immediately. Our song and theirs are one. We listen to their melody—their soul, he says—and "write down the notes without diluting or de-praving them."[13] We allow ourselves to be addressed and drawn out.

Dewey writes of "being lured on by an aura."[14] This can only mean that sensuous presences—typically subliminal and unacknowledgeable by prosaic consciousness—already envelop and solicit us, pulling us into what will intensify and clarify them. Art's truth is just the fuller realization of where we already tend as ancient bodily beings caught up in a world in which meaning is made—the realization of a felt inevita-bility. Meaning as truth is a benevolent spell—a call to return home to the awesome world that formed us. Circular power returns into itself.

No accident that art and religion arose together, and might even today feel their way back together as so many grope for regenerative experience.[15]

Art and Hunger for the Earth

As Gaston Bachelard shows in *Water and Dreams*, without a sensuous and imaginative involvement in matter, our lives lack substance. Nature acknowledged is a sustaining "directly valorized reality."[16]

Bachelard re-evokes the four archaic elements, earth, air, fire, and water.[17] He recalls Hesiod's ancient warning, "Never urinate at the mouth of rivers which flow into the sea, nor at their source: be very careful about this."[18] Hesiod is enunciating "a single rule of primitive morality (that) defends . . . the maternity of the waters." Urinating di-rectly in water's source short-circuits Nature's cyclicity. Water's powers of regeneration are self-evidently valuable, and art's truth is a chief way of disclosing and radiating this value.

The ancient idea of the four elements exhibits human experience at a point before scientific and artistic investigations diverged, a parting that leaves art in the shadows of "the merely subjective," shorn even of the title of investigation. And science and technology isolated from art are left without clearly recognizable rootage in the fertile ground of meaning and the maturation of our lives. Loneliness and desiccation have become unspeakable reality. Bachelard writes,

> Who . . . does not feel a special irrational, unconscious, direct repugnance for a dirty river? . . . the *unconscious value* [that is] attached to pure water. In the threat to pure crystalline water, we can measure the fervor with which we greet, in their freshness and youth, the stream, the spring, the river, and all reservoirs of natural limpidity. We feel that metaphors of limpidity and freshness will be secure as soon as they become attached to such directly valorized realities.[19]

Wonders of scientific investigation dominate the popular mentality; reductionist ways of casting meaning, truth, reality conceal or trivialize the core of our experience. So it is commonly thought, "The river is really only certain fluid molecules flowing at such and such a volume per second, etc. We *say* 'sobbing' of the river, predicate it of the river, but it is not a property of the river. Its 'sobbing' is only what we make of it within our minds."

On the contrary, as we actually experience the river it is sobbing. Let us say, paint, sing it. Why live stunned, not even attempting to articulate loneliness and grief? Why care if some think us stupid because they know we can't squint down and measure and weigh and time the sobbing?

Contemporary materialism and its correspondence theory of truth prompt some to snigger, "But who is to say what value is? The industrialist who sees the river as an efficient sewer for carrying off his factory's wastes will see 'the filth' of the river as a value. He might chuckle with delight."

But if Bachelard is right about the primitive morality of directly valorized realities, and right about the universality of such sustaining and consolidating experiences of intrinsic value—and I think he is—the hypothetical industrialist's ecstasy is constricting even for himself. But, even if we grant him a kind of delight, it cannot be allowed to dispirit us and to ruin the health of those who must try to live with his handiwork.

Bachelard's poetic prose shows crystalline water purifying and rejuvenating in fact—but not in a way that "factual inquiries" could discern. "Running water, gushing water, is primitively *living water*. . . . The psychology of purification is dependent on material imagination and not on external experience."[20] In a similar vein, we know that free waters—wild rivers—find their way back to the sea, opening onto the sea's horizon, finding their way to the source of all living things on the planet. So "Water in its symbolism can bring everything together. . . . A great poet imagines intuitively the values that belong naturally to a profound life."[21]

The truths of art must complement those of science. Correspondence truths in the form of accurate predications of properties to substances existing independently of us—no matter how startling and significant—are not sufficient. The names and images we need are those that elicit our sobbing or laughing with the rest of the world. The last lines of the book:

> Come, oh my friends, on a clear morning to sing the stream's vowels! Where is our first suffering? We have hesitated to say. . . . It was born in the hours when we have hoarded within us things left unsaid. Even so, the stream will teach you to speak. . . . Not a moment will pass without repeating some lovely round word that rolls over the stones.

Music

Music is perhaps the greatest art. Melodies give us the souls of things, said Emerson, and Bachelard prompts us to sing the stream's vowels. In the artistic rendering of the world, the serpentine body's sounds and the stream's are ecstatically one. Each coils into and instantly confirms the other within the resounding world that feeds back into itself.

When the stream coaxes us to repeat "some lovely round word that rolls over the stones," both the linear and cyclical dimensions of the enduring World-whole are woven together, both passing time and the eternal present, and we are woven whole as distinctive nodes within it all. Ilya Prigogine and Isabelle Stengers, scientist and philosopher, write:

> It is hard to avoid the impression that the distinction between what exists in time, what is irreversible, and, on the other hand, what is outside of time, what is eternal, is at the origin of human symbolic activity. Perhaps this is especially so in artistic activity.[22]

Music in the deepest sense animates all the arts: sculpture, painting, dance and theatre, poetry. For, stretching through time, the underlying rhythmic reciprocity of body and world, essentially musical, orders, arrays, collects, and re-collects all things sensuously—meaningfully.

There is no culture without music, because humans need its continuity, rhythm, momentum to abide the shocks of life moment by moment: those irruptive cravings, impulses, and fears that open up black holes in time.

Emerson says,

> We are lovers of rhyme and return, period and musical reflection. The babe is lulled to sleep by the nurse's song. Sailors can work better for their *yo-heave-o*. . . . Metre begins with pulse-beat, and the length of lines in songs and poems is determined by the inhalation and exhalation of the lungs. If you hum or whistle the rhythm of the common English metres . . . you can easily believe these metres to be organic, derived from the human pulse, and to be therefore not proper to one nation, but to mankind. I think that you will . . . be set on searching for the words that can rightly fill these vacant beats.[23]

But perhaps it is wordless music that lures us most silkenly out of conventional orientations and constrictions as well as momentary disruptive impulses—that spooling uncoiling that reveals inner wilderness body interweaving with outer wilderness world. Such music is of that basal order of meaning that cannot be paraphrased, translated, or put into any other terms. The sailors' *yo-heave-o* is more sounds than words, and no mere expression, pressing out, of an "internal mental state," but rather a releasing of sighs and sobs and groans into the rippling and rumbling waves against which the sailors bend their oars. When this release and the water's response take the form of repeatable articulation, we have art's truth, which is intimate with ritual.

Thoreau writes,

> All good things are wild and free. There is something in a strain of music, whether produced by an instrument or by the human voice—take the sound of a bugle in a summer night, for instance—which by its wildness . . . reminds me of the cries emitted by wild beasts in their native forests. It is so much of their wildness as I can understand.[24]

Our eyes were closed in the womb, but after a certain point we could hear. Muffled sounds, basal rhythms and novelties coming from outside the mother's body interwove with basal rhythms and novelties from within her. And both these with sounds from within the unborn infant's body, the beating of its own heart. And all in the darkness. This is the primal generative and regenerative wilderness for each of us—will-of-the-place, *Contact!*

There is no culture without music, for no infant can emerge in the outer world without already being tuned moodily in various ways. Each comes to feel at home in the world and be whole only when attuned to the Whole through each one's various situations, and this, again, is the general principle of value: the greatest coherence or unity within the greatest variety and individuation. There is linear time, change and particularity, but also shared patterns of fecundity that repeat themselves within the regenerative Whole and carry each of us in one way or another. Prosaic truth limps lamely after the truth that we *are*—particular and fused.[25]

Art and Religion

Paul writes in Romans 8:26: "But the Spirit itself maketh intercession for us with groanings that cannot be expressed in speech." The passage suggests trance states of angelic possession, when we are so attuned to the world that, in a sense, we needn't do a thing.[26] The clouds buoy and sustain and the stream leads us on sacramentally. To be regenerated is to be set resonating spontaneously by the world. To create in art is to invite and amplify this resonance. It is music, most obviously, that excites horizons of resonance within the fertile darkness of body-self.

We echolocate, we come back to ourselves from the resonant world. We hear not just the sounds themselves—whatever that might mean—but hear them as hearable by others, a vast chorus of fellow beings, known and unknown. We hear songs of streams and insects, sighings of trees and other beings, lamentations, rejoicings, all the voices raised to Heaven and sent reverberating into answering Earth. We experience our experiencing experienced by known and unknown others who come to find us. William James:

> Music gives us . . . a verge of the mind . . . and whispers therefrom mingle with the operations of our understanding, even as the waters of the infinite ocean send their waves to break among the pebbles that lie upon our shores.[27]

Dewey writes that art is more moral than moralities.[28] For art bonds us ecstatically with the rest of the vast world, a power that lies beyond the reach of prosaic truth.

When a graduate student, I had a photograph of John Dewey on the wall of my room. It showed him in later years—face sagging and deeply lined, a shock of gray, unruly hair almost pasted to his forehead, his eyes watery, dim, faded.

Once when she was on a visit, my mother noticed the photograph and asked who he was. I said it was John Dewey, the famous philosopher. She said, "He looks like the greengrocer to me," and left the room.

Chagrin vied with contempt. But now I see how acute her insight was. Yes, he does look like the greengrocer. He wanted to nourish us. He wanted to restore effective contact with Nature which we were on the verge of paving over and forgetting. In the midst of affluence, we are starving, he thought. I must have seen Dewey looking out at us like the "green men" do through their stony leaves on cathedral pillars—vestiges of ancient pagan times trembling with possibilities.

We do not gain entrance to Nature only through the keyhole of science and the powerful key of technology. Outside the focus of consciousness and facile speech we are enwombed with her forever. Some ways of living are conducive to growth and connectedness; others lead to self-contradiction, stagnation, and addiction.

Perhaps it is ridiculous to suppose that personally and corporately we might develop an art of life in which ecstatic projects interweave themselves within circular power returning into itself. Perhaps preposterous to imagine that cultures might resound again within the sacred, that all things, belonging together within the ever evolving and integrating matrix, might resonate awesomely through ourselves. Might we find something like our own Dreaming?

Notes

1. See Catherine Keller, *From a Broken Web.*
2. *Blood Relations*, 43, 48.
3. *Experience and Nature*, 295.
4. Knight, 46.
5. Ingmar Bergman has said, "I regard art (and not only the art of the cinema) as lacking importance in our time: art no longer has the power and possibility to influence the development of our lives. . . . On the whole, art is free, shameless, irresponsible: the movement is intense, almost feverish, it resembles . . . a snakeskin full of ants. The snake is long since dead, eaten out from within, deprived of its poison, but the skin moves, filled with meddlesome life." (Quoted in Paisley Livingston, *Ingmar Bergman and the Rituals of Art*, Ithaca and London, 1982, p. 14. I am indebted to Harry Redner for this reference.) This is profound, but I don't believe it's the last word.
6. *Art as Experience*, 24.
7. 104.
8. Concerning the absence of any blackboard upon which replicable images

could be compared to objects "out there," see Wolfgang Pauli, *Writings on Physics and Philosophy*, 260. He does not use the image of a blackboard, but he implies the general idea.

9. I am indebted to Hilary Putnam for suggesting such a wording.

10. When the radical right denounces news commentators and pundits as "liberals," they are probably objecting (however inarticulately) to unquestioned secularism and scientism.

11. Quoted in Elting Morrison's review of Heisenberg's *Physics and Beyond, New York Times Book Review*, January 17, 1971. Wolfgang Pauli writes of "the ever present unconscious and instinctive ingredients of thinking" (*Writings*, 128), and of "the metaphorical preliminary stage guiding the instinctus" (229, 232).

12. Intent upon extirpating myth root and branch, many today would say that the meaning of "The dog is in the garage" is just the conditions that would render that statement true if it were to be true. But only prosaic meaning is reducible to truth conditions. To ask, "Is the proposition 'The dog is a mythic creature' true or false?" is doltish. Mythic meanings are unconditioned or absolute—such as World-Whole or God—and elude specification of truth conditions. How could we ever know we had nailed down the conditions that would verify assertions about these? Or, do I *really* sense the "everything else beyond the horizon"? Should I try to prove I know its truth conditions? A serious response to an obtuse question is obtuse.

13. "The Poet," 273.

14. *Art as Experience*, 73. On 349 he writes, "Art is a mode of prediction not found in charts and statistics."

15. The roots of "religion" = *re-ligere*, to be tied-back (to sources).

16. *Water and Dreams*, 137.

17. 3.

18. 135.

19. 137.

20. 141.

21. 149.

22. *Order Out of Chaos*, 312.

23. The love of "rhyme and return" suggests Circular power returning into itself regeneratively. See Emerson's "Poetry and Imagination," in *Ralph Waldo Emerson*, ed. Richard Poirier, New York, 1990, 459–60.

24. "Walking," 34, 1994.

25. Nevertheless, scientific truth in our century sometimes converges to form a superaesthetic truth. For example, Paul Robertson: "At the University of California at Irvine, researchers mathematically modelling the brain successfully translated neuron firing patterns into sound and heard music. . . . Once the algorithmic 'neural' patterns were displayed as auditory information, the repeating and changing patterns seemed curiously like tonal classical music." *BBC Music Magazine*, Summer Special, 1996, 24. Concerning the need to interconnect science, art, religion, see Wolfgang Pauli (*Writings*, 139–140): "I believe that it is the destiny of the occident continually to keep bringing into connection . . . these two fundamental attitudes . . . the

rational-critical, which seeks to understand, and . . . the mystic irrational, which looks for the redeeming experience of oneness. *Both* attitudes will always reside in the human soul, and each will always carry the other already within itself as the germ of its contrary. Thus there arises a sort of dialectical process, of which we do not know whither it is leading us. I believe that . . . we have to commit ourselves to this process, and recognize the pair of opposites as complementary." This I applaud, but have reservations about the locution "mystic irrational."

26. Note also Jonathan Edwards, "I often used to sit and view all nature, to behold the sweet glory of God in these things; in the mean time, *singing* forth, with a low voice my contemplations of the Creator and Redeemer. . . . it was always my manner . . . to sing forth my contemplations. I was almost constantly in ejaculatory prayer" (*Devotions*, 5).

27. *Varieties of Religious Experience*, "Mysticism," my anthology, 247.

28. *Art as Experience*, 348.

PART III

Harmony
with
Nature

It is possible to take the female genitalic outlines for what they are: vulvas give entrance to miraculous insides. From them issue both blood and tiny living beings. A vulval shape depicts a bodily opening from which emanates *extraordinary* substances and forms. Males are not the source of such extraordinary happenings; males have no such *insides*.

—MAXINE SHEETS-JOHNSTONE

It is a question of finding in the present, the flesh of the world . . . an "ever new" and "always the same" . . . Existential eternity. The indestructible. . . . It is the flesh, the mother.

. . . it is the painter to whom the things of the world give birth by a sort of concentration or coming-to-itself of the visible.

—MAURICE MERLEAU-PONTY

How can a man be born when he is old? Can he enter the second time into his mother's womb and be born?

—JOHN 3:4

Chapter Eleven

Mother Nature:

Circular Power

Returning into Itself

The original mother-child greeting ritual is the base on which sacred ritual is built.

—DOLORES LACHAPELLE, *SACRED LAND,*
SACRED SEX, RAPTURE OF THE DEEP

Some years past, I was driving up the coast of California from Los Angeles, back to origins. Near Ventura I crossed over Sespi Creek. We were teenagers here on a fishing trip about forty years ago. The best fisherman among us had forgotten his fishing pole. There was oil scum on the tails of pools, and fishing was poor.

But now something motionless in the sky fixed attention. How could it just hang there? Its immense wings rigid and motionless—a condor. Uncanny, hard to believe. Something beyond the boundary, yet in full presence. Too big to be real, but it was. For a few moments it was visible, then the great bird canted its wings slightly, floated slowly across the sky and out of sight behind a mountain.

No need to get out of my car now and look at the empty sky. I continued up the coast, beautiful as the bluish mountains turned brown closer by and fell into the sea. The line of the horizon was visible, slightly misty, but there. *Beyond it lay everything else—the universe.*

Then near Santa Barbara the oil-drilling platforms began. Standing on their multiple pilings, several miles out from shore, they were not close together. Still, they formed a fence that cut me off from that line. I imagined standing on one of them and looking out to sea, but it did not help. I turned away and looked at the mountains.

I could not remember not seeing mountains, particularly those bounding Los Angeles on the north and east: the San Gabriel Mountains, named for the archangel Gabriel. But the older name, Sierra Madre, Mother Mountains, was better. In the '30s and '40s they were etched against the blue sky, visible nearly every day from almost any point in the city. Every roll, turn, shape was as familiar as breath, a huge reclining body that pulled me softly into itself. Sierra Madre.

Almost as familiar were the mountains in the desert beyond Cajon Pass, standing high to the south of our small house in Lucerne Valley, the Big Bear Mountains. The capital peak was directly behind us, its whitish face grooved by rain and wind through the millennia. I imagined that even here in the desert water would gather at its base after a storm—just over the hillock that cut off vision. In her artist's hand, my mother had lettered above the large window opening onto the peak: "I will lift up mine eyes unto the hills, from whence cometh my strength"—a Psalm of David.

The whitish look in these mountains was the presence of lime. The mining of it began further down the range. Over the years the road on one of the peaks switchbacked ever higher, affording access to the shafts and allowing the mineral to be trucked away. At night the lime works glowed just over the swale, the near horizon. After a year or two the road was extended to the top of the mountain and left a gash, as if a creature's head had been cut off. After twenty or so years the road reached the capital peak and the burrowing and cutting began at its base; it now continues up its face. The operation is the work of Mitsubishi Corporation of Japan.

Sometimes going from Los Angeles to New York, I fly over this area. Two blotches of lime, each about 40 square miles, cover the valley floor. The mineral is piled and processed at two points, but the strong winds blow it here and there on the face of the desert. It is essential for the cement that makes highways, sidewalks, shopping malls, apartment houses, swimming pools.

Spread now through the world, North Atlantic influence exhibits symptoms of massive corporate addiction. Insatiable hungers seem normal.

The International Growth Society accelerates. The momentary "rush" of profit sheets masks out the longer range degradations and depletions of environments and persons. To keep the juggernaut going, ever more rushes are necessary.

Technological-commercial expansion is addictive "adaptation," an artificial state. It is the appearance of control, floated by rushes (as long as they are available), and by draining resources and social organs to the point of imminent, but still ignorable, collapse. Like addicts, the corporate individual craves what will finally destroy it.

Our hunter-gatherer minding bodies have great difficulty realizing that we have degraded Nature herself. Though cultural and technological evolution is much more rapid than biological or geological, it is still too slow to capture attention. So we are distracted by the latest news, the dramatic and sensational event, the ultimate upshot of which tends to be vanishingly small. In the midst of a world radically altered by humans over a few centuries, our ancient eyes still turn to whatever moves abruptly in the immediate field of vision—movie and TV images, for example.

We have not learned to live wisely, as organisms who transform Nature into culture, but always within Nature. How do we remarry Nature and culture? If the mythos of the Mother were to be regrafted somehow into our living, an ecstasy that is quietly at home with all beings—past, present, and to come—might be reborn.

Mother Nature and Defiant Male Identity
The most important suggestion is greatly difficult: To stop habitually modeling ourselves on a male deity that stands above Nature and owns and controls her. Probably this pattern has become ingrained in nervous systems through roughly six thousand years of belief and behavior. Only through reconditioning ourselves bodily can the patriarchal pattern be counteracted. The whole body must take the lead, but with the enculturated big brain now chastened. Perhaps we can revive the most ancient myth we have of ourselves and Nature: Mother Nature enwombing us within herself. If so, we will better own up to our own bodies' primal needs and urges, better allow body-self to be true to itself, less apt to repeat gratifications addictively.

Aside from nutrition and shelter, the male's deepest need is to establish an identity different from the mother's. From the beginning, the male child is sexually other than his birthing source. This otherness is clutched as a ground for identity within the tremulous zone of primal body-self. But without the ground shared with mother, the male would have no being at all.

Within recorded history "more advanced" cultures, culminating in our own, have shown the strains of male assertiveness. "I am self-sufficient, independent of mother and the bonds of matter." The paradigmatic maker, doer, knower, has been God-the-Father—whom Christians and Muslims characterize as a pure transcendent Mind or Word (*Logos*). In forming identity humans have modeled themselves upon Him. Males tend to separate from Mother Nature, trying to detach from circular power returning into itself. Male assertiveness becomes addictively aggressive. For, continually denying what we continue to need—maternal energies of all sorts—we can never get enough of the assertiveness we think compensates for the loss. Embodied now in international corporations, male addictive assertiveness is destroying Nature's regenerative systems.

Most male myths are not content to say that the male child's sexual difference from the mother is one element in his identity, but define the male in separation from her. Likewise, a sharp line is drawn between the strongest male and everything else. Thus generated is the hierarchical model ingrained in the corporate body and its members: the master-mind dominates and controls all that falls beneath him. Male bias emphasizes difference at the expense of what is shared, conceals what males do share with others—including male bias itself. This is patriarchy.[1]

Male Aggression Against Women

The most basic of all needs is to bond initially with the mother.[2] When this is not satisfied, fear and blind rage toward her rouse, and some mad attempt to bond with her may ensue. As Joseph Pearce pointed out, this is sometimes rape.

Pearce recounts two cases in which women, threatened with rape and death, somehow discerned their attacker's needs, recognized them calmly, and ministered to them on the spot. The attackers desisted. The neediness and precariousness of human development, especially males' now that viable rites of passage and initiation have disintegrated, help explain these incidents.

In the first case, the woman saw how frightened her attackers were, and, having resigned herself to death, asked solicitously about the reasons for their fear. They responded that she must answer the question they had earlier posed: How does it feel to be about to die? She collected herself, not mirroring and reinforcing their hysteria ("womb sickness"), and resumed her inquiries into their fear. As one of the men attempted to mount her, he collapsed in sobs, and the other pounded on the ground in grief:

> "What is it? What is it? What's gone wrong?" Then he, too, burst into the same strange, grief-stricken sobs. . . . It was some time before they quieted enough that she could speak to them and say quietly, "Boys, we may as well go home." Without a word, only their continued sobbing, they drove her back to the city. At the first subway, she suggested they let her out, which they did. She . . . had $300 in her purse, but they had given no thought to money. On impulse, she asked them would they lend her the money for the subway, which they did. She turned her back to them, started down the steps, heard them drive away, put her money in the turnstile, walked through, and fainted dead away. When she was questioned by the policeman who revived her, she replied, "If I told you, you'd never believe me."[3]

The other case is similar. Awakened in the dead of night in a remote Kansas farmhouse many years ago, a grandmother, who had been asleep in bed with her granddaughter, confronted a huge man,

> . . . rain dripping from his chin, poised over them, a large chunk of firewood in his hand, raised high . . . The grandmother [said] to the man, "I'm glad you found our house. You've come to the right place. The door was unlocked for you. You are welcomed here. It's a bad night to be out. You are cold, wet, and hungry. Take the firewood you have there, go in, and stir up the kitchen stove. Let me get a wrap on, and I will find you some dry clothes, fix you a good hot meal and a pallet behind the stove to sleep, where it's good and warm". . . . She said no more but calmly waited. A long pause ensued. Slowly the stick lowered, and the man said his only words the entire visit: "I won't hurt you."

She put the man up for the night, but he vanished before dawn. The sheriff appeared in the morning with bloodhounds. The man was an escaped homicidal maniac who earlier that evening had killed a family.

Pearce writes, the story was "related to us by a dear elderly woman [the granddaughter in the account of long ago]. I can vouch for its authenticity in spite of its grade-C movie quality."[4]

Identity and the Male and Female Bodies
Thoughtful people within Western patriarchal "high religions"— Judaism, Christianity, Islam—are challenged: to maintain the rooting

continuity of local custom and loyalty, but also to open out into an emerging world-consciousness in which interdependency is acknowledged and authority shared in a communitarian, nonhierarchical, and, I think, matristic way.[6] A new kind of leap of faith is required, one that does not try to transcend, suppress, or manage primal needs. For this forces them into addictive substitute gratifications that are, ironically, uncontrollable by the will.

The patriarchal frame of mind and its archetypes picture identity not in terms of how we are included in the whole—circular power returning into itself—but only of how we exclude things, often brutally.[6]

The man's body presents itself as hard, firmly enclosed, nearly self-sufficient, uninteresting in its interior, sharply different from woman's. It cannot generate another individual within itself, or even allow another individual within itself.[7] When this body poses as the archetype of humanity, we tend to conclude that any being is individual because it excludes others. So the world appears to be merely an aggregate of allegedly self-sufficient atoms. The dominant tendencies in Western philosophy and patriarchy have been sharply individualistic. They foster terror that our boundaries can be breached. Denying vulnerability and needs prompts systematic terror to keep others at bay—indeed, systematic terror to keep all needs at bay, or suppressed, a pervasive paranoia.

Another view of the world emerges when we believe the female body also to be paradigmatically human, when we look through the female "lens." A woman has the capacity to generate another individual within the envelope of her body. Using her as model or archetype, we are less apt to think the envelope defines an ultimate unit of reality—that her individuality requires the exclusion of other beings from it. We better understand the ecstatic experience of being both self and other. Beings-in-community tend to become fundamental.[8] Emerson's circular power reappears, and in a form fuller than he realized, burdened as we was with some sexist bias.[9]

If it is true that at least in general and over large populations the female body tends to individuate itself somewhat differently than the male does, this has the greatest significance for understanding knowledge itself. Since knowing is, roughly, a matter of processing the world through aware bodies, a body-self that individuates itself in a certain way tends to become the template for grasping individuation everywhere. Perhaps certain generative (or otherwise relational) properties, may be best grasped and perceived by women, on average, than by men.[10]

Evidence begins to appear that hormonal differences between the sexes fashion their brains differently to some extent.[11] "Different from men therefore inferior," is a fallacious inference deriving from the unwarranted assumption that males are the norm and anything that deviates from that is deficient. We are all needy body-selves.[12]

Knowing does not happen "inside a mind," in some matter-and-gender-neutral zone. Knowing, and the truth it leaves us, is more or less intimate intercourse of organism and environment, and organisms are sexually differentiated. As Dewey discovered, the female archetype, emerging without eliminating the patriarchal heritage, rounds out and balances thought, feeling, and action, and provides a regenerative matrix within which our many ecstatic impulses and abilities to know and to be can be fostered.[13] To metaphors for knowing such as "lens," we should add "feeling and accommodating through membranes."

Above all, employing the female archetype we are less apt to think that civilization excludes profusive life of wilderness, and more apt to mediate the boundary between wilderness and civilization and to engender vital life within will-of-the-place. This is what the hags, the witches, did, and excited the terror and rage of the men who killed them.

Why Myths?

To ask why we need myths—archetypes such as female—is like asking why we need to be human. Whether acknowledged or not, myth tellings and harkenings inform our identities, which tacitly model what can be known. Various positive archetypes are excited valuations and meanings that impart an eagerness and direction for life. The past does not disintegrate and vanish. It blurs into the present, haunting with mythic archetypal figures. Technology is a threat today because it transpires within a mythic attitude toward the world that is fairly recent, patriarchal—fixed unblinkingly on separation, manipulation, and domination. Primal needs are defied whenever they block the road of ambition.

An archetypal way of dealing with the world such as female is not just "thought up." It was evolved over millennia through that funded, time-proven interaction of humans with the rest of Nature that *is* culture ("culture"—where things grow). First to be considered are the distinctive reproductive capacities of male and female body-selves. But these aren't isolated in an exclusively biological or material domain. To see them as such is to be blinkered by a dualism of mind (or culture) over matter and, in this case, to emphasize the "lower physical side" of it. To speak of *sex* differences is to stay within the terminology of Western biology.

We must speak of *gender* differences. Archetypes, basic models of living, can't be grasped dualistically as simply biological or simply cultural. And archetypes can't be grasped through mechanistic and linear understanding as either simple causes or simple effects. Speaking of male and female archetypes is finding ourselves in the broader and fuzzier area of gender differences. They arise over time at the deepest and murkiest intersection of biology and culture. Even human children born at full term have only partially developed brains. These continue to develop as a function of the child's experience in human environments. Every culture gets in on the "wiring" of the brain in somewhat different ways.[14]

To assume, however, that gender formation is "simply cultural" is to be caught up in the patriarchal opposition of mind or culture over against Nature or matter, but to emphasize this time the "higher cultural" or mental "side." All change must begin where prehuman and human biocultural development has brought us to this moment. It is a fact that females have two X chromosomes and males an X and a Y, and likewise that these differences typically entail, for example, hormonal differences that seem even to fashion brains differentially. Not to honor the relatively stable aspects of the world is to weight eccentrically the precarious ones.[15]

The very distinction between male and female archetypes may perpetuate patriarchal tendencies to split the world into oppositional dualisms. And yet, and yet, the distinction cannot simply be thrown out either; its vast inertia persists, and some gender distinctions must be made.

Some feminist thinkers have written of creating "multiple genders," and of the need to "reinvent gender."[16] Given the rapidly changing world scene, I agree. But we must first know where we are if we are to begin changing effectively. Archetypal dualism has in fact variously formed world cultures, including our own over many millennia. There is no accounting for our addiction prone, Nature-defying culture without assuming the dominance of the Male archetype.

The Persistence of Archetypes

Male and female archetypes of being and behavior are still present in our culture, but the rapid flux of events wears away their edges. Biocult-

ural evolution moves us quickly, incalculably. Take gendered sexual signaling, for example. As Maxine Sheets-Johnstone has well shown, in
sexual signals we find an amalgam of inherited behaviors, both human
and non-human, still present despite cultural change.[17] For example,
penile and vulval displays in sexual courtship, rooted in prehuman pri-
mates, are, for humans, no longer entrained in biologically rigid oestral
cycles. Human displays are regulated in culture-induced and policed
patterns that have themselves become archetypal. Take the pan-human
prohibition against women opening their legs and exposing their genitals, except at appointed times and places deeply ritualized. Sheets-
Johnstone writes, "If we ask how there can be a corporeal archetype
when its regular enactment is not in evidence, we find the rather
obvious answer that cultural overlays mask the archetype; exposure-
avoidance is precisely the cultural mask."[18] There are taboos against
exposure only because of a residual tendency to expose, perhaps a need
to do so that should be honored even if not acted upon. Bullishly denying needs creates addictions.

The common "culturalist" mistake today is to suppose that the prehuman stratum of biologically regulated sexual display can simply be
stamped out, or that it will wither away on its own. But there still are
anciently rooted human sexual displays to attract mates, and the needs
so to display.[19] In heavily patriarchal societies, men's fear of loss of
control generally reduces the complex repertoire of women's sexual
communications. Women are merely subservient "holes to be filled" by
the "active, intelligently organizing male." But without parity, without
feeding into each other reciprocally, there is addictive forcing of gratification.

Organically Developing Archetypes of
Knowing and Being
Traditionally, male and female body-selves have had different functions
throughout life. Over myriad formative years, females in gatherer-
hunter societies provided the inner circle of the community, hunting
for smaller game than did males, gathering, and engaging in their own
rites of passage and crystallization of identity. Males departed from this
circle in their surveying and hunting expeditions and returned to it in
due course. These ways are long past, but they formed our nervous and
hormonal systems.[20]

All innovation occurs in the middle of things, at a given stage and
moment of biocultural development. To forget, for example, that the

male body-self simply does not have the capacity to conceive and give birth, or to forget that even the most nurturing father will not lactate when the baby cries, is counter-productive.[21]

Maintaining a vital life over a lifetime, I imagine, will require those reinventing their gender to patiently do body-work—some of it perhaps through ancient **spiritual disciplines. The painstaking work of "rewiring" nervous systems in indefinitely many ways must be done, otherwise patriarchy will hang on addictively as the dominance of the mental or the cultural over the "merely physical." To be chronically uncomfortable and dissatisfied within Nature invites burnout and addiction.

What is vital does not simply defy the earlier life out of which it grows. For example, in only the past few decades in this century American women's way of being athletic has altered dramatically.[22] But this need not involve contempt for "feminine" predecessors in earlier centuries, and might remind us of ancient women ancestors hunting and gathering and transporting heavy loads.

Entrenched Patriarchal Myths

In a culture dominated for thousands of years now by distinctly male attitudes and archetypes, male daring, "self-sufficiency," and cruelty pulse and shine with numinous power. Certain evolutionary theorists trace much of this to male sexual impulsivity and opportunism stemming from their nil burden in pregnancy. The theory may also help explain the dominantly male technological and commercial rampage across the planet.[23]

Myths of the dominating male, coupled with his greater physical power, tend to dominate everything, including other myths. Isn't it time to deconstruct them?

Easier announced than carried out, for male myths capture thought and energy. Attempts to displace the hierarchical pyramid tend to mimic it. "Just identify the problem and attack it!" That is the male solution. At work is the dominating archetype—the supreme male doer and thinker—limiting imagination to violent forms of intervention. But that is the real problem, so of course it cannot be seen. It allows no time for dilating alertly, letting things be what they are within the sustaining matrix.[24]

In forms preserved in patriarchy, most Mother myths equate the female with mere reproductivity, and limit her to supplying only the matter and housing of the child. (The roots of *patriarch* are entangled with *pattern*, and *mother* has roots in *mater* and *matter*; the pattern is imposed on the matter.) Until a few hundred years ago, learned people thought that the father was the true parent: through insemination, he implanted the human form of the new individual ("rational animal") into the receptacle of the womb and its matter. The effects of this linger in the culture. Thus the addictive raping of Nature is rationalized within the male paradigm or archetype that it takes completely for granted.

Notice how understanding of the Mother occurs within Father mythology. In accepting the idea of the female as simply the passive opposite of the active male, typical thinking is perpetuated: male sundered from female and superior to female; mind sundered from matter and superior to it; activity sundered from passivity and superior to it; parts sundered from wholes and wholes merely aggregates of parts. There is no reciprocity, organicity. This kind of thinking should be understood as a failure to abide by the general principle of value: sound individuation and differentiation because of vital membership in the World-whole.

If patriarchal archetypes' hobbling dualisms are not unlocked, addictions will continue to proliferate. What if neither active nor passive best describes human beings when most vitally involved in the world? Just as neither body nor self does, and we need a third alternative— body-self—so neither active nor passive, neither in control nor out of control, but poised, joyous, and spontaneous doing-and-allowing. If we patiently developed new habits, and did not force feed-back from Nature, we might rediscover the meaning of it as evolving and rooted mother.

We can try to place the dualizing patriarchal myths within the more inclusive field of the Female. After all, in the experience of pregnancy a woman is, in a way patriarchy finds excruciatingly hard to understand, both self and other.[25] Our very myths of regeneration must be regenerated.

I had not visited the Sierras for eighteen years. Living in New York City, "making it," establishing a career as professor, marrying and fathering two children—I had allowed all that to dominate my life. Now the Sierras came home to me, charged through the intimate new group of which I was a member. Yet I wanted to be alone for a few moments.

The family was getting the evening meal in our VW Camper. We had parked in a clearing at about 9000 feet, along with ten or so other recreational vehicles. Dusk settled upon us and I slipped away to the long-branched bushes that typi-

*cally line the Sierra streams at this altitude—Bishop Creek in this case. I stopped
to peel some branches from my face.*

*With the softest murmurings at my feet it greeted me. There it was. How
silent comes the water around that bend. It was a streaming, uncoiling turning
toward me that seemed to come from the everything else, the vast universe itself
beyond this spot. The stream snaked, rolled, and rippled, merging rainbowlike
within itself, from transparent and crystalline at the banks to turquoise and dark
blue in its depths. The constrictions of my life and its frenzy dropped away as if
cured by miraculous intervention. For a few moments the coursing rhythms of
the stream were the coursing rhythms of my life.*

If fear could be dissolved for a moment, the Great Mother might be felt
to be implicit in all things. She is Earth Mother, and earthlings, regardless of gender difference, participate in a power and spontaneity that
vastly exceed any of our voluntary careers.[26] Since all living and non-
living things share her womb, all are kin. She is mythic fuel for cosmic
democracy, "live and let live." Most importantly, the Mother archetype,
ingrained in our bodies, inclines us to experience bodily needs and
desires as integral to our very selves.

The Father archetype is not eliminated in this theory, only removed
from the apex of a hierarchy and placed in the ever-regenerating and
evolving Whole, Circular power returning into itself. The Father or
male archetype, deep-dyed in hunting and exploration, points beyond
the easily and routinely known, but cannot point beyond the Whole,
and our never completely eliminable limitations within it. Trying to do
so is the addictive attempt to transcend Nature.

Grammatically, "Earth Mother," "Great Mother," or "Mother Nature" are singulars. But we needn't think of a single, separate entity.
Thinking within the female archetype we conceive of a cycling *web* of
nurturing beings—including the two-legged, the four-legged, the
winged, the swimmers, the crawlers. The female archetype helps bridge
the gulf between indigenous, shamanic cultures and the European tradition.[27] Moreover, it facilitates an understanding of how persons other
than the biological mother can play the motherly role: they are caught
up in the funded power of the Mother archetype, and there's no pre-
dicting exactly how a particular individual will do.[28]

The female and male archetypes engender different forms of ecstasy. The former is that of the self-recharging circuit or Circle. Body-
self shares itself and returns to itself augmented by the responses of
others. Male ecstasy resembles the pyramid: the great ego at the apex

prevails for awhile, typically at the expense of beings beneath him. But when only obsequiousness and agony is fed back into the master from subservient others, he lacks the fullness of the Whole, and that trust in the Whole that sustains, that carries through black holes in time, and finally prepares for death. Tyranny is a particularly restless and finally draining trance and addiction.[29]

To assume that truth can only be a property of prosaic literal statements asserted from the commanding "objective standpoint," of "asserted propositional content" is a form of patriarchal tyranny. Words that point only to what is taken to be an "objective situation" cut through and ignore the context in which individuals can intimately participate, where no prosaic, easily communicable truth is adequate.[30]

Women's Religions and Healing

Susan Starr Sered has compiled voluminous research on extant women-centered religions worldwide. Her findings bear out the general overview of gender differences here presented. Women's religions tend to be much more communal, egalitarian, pluralistic, less concerned with finding a single commanding viewpoint from which the whole can be surveyed and issues adjudicated through a sweeping theory that purports to get at *the* objective truth.[31]

Not surprisingly, in women's religions feelings are allowed cognitive status. The patriarchal split between "feeling body" and "knowing, commanding mind" is rejected. For how can we know what our needs are unless we know them feelingly? Sered points out that in female-centered healing ceremonies the first step is to acknowledge that the sick woman's suffering is real. Then in the most intimate reciprocity between group, client, and woman-shaman, the client's needs are elicited. In a world formed over millions of years regeneratively, needs call out to what can satisfy them—if we are listening.

The relevance of Sered's findings to the plague of addictions is clear: If addiction springs mainly from the failure to clearly identify primal needs, then the addictive flailing that tries to satisfy them is accounted for, and may be alleviated through a feeling-knowing of the needs.

As Sered and others make clear, healing ceremonies are communal, and carry the weight of the group's authority and history. What are we to do in today's fractured times, when, as Eric Erikson points out, we have "elderlies" but no "elders."[32] And when science and technology no longer cohere with folk traditions and fine arts? We can only search for intrinsically valuable experiences that carry authority.

Male and Female Feed into Each Other

Certain indigenous cultures are free of the rigid oppositions of male and female, and the simplistic idea that men are only male, and women only female. In the tradition of *couvade* the male ritualistically simulates birth pangs, often winding up exhausted, the newborn infant placed on his sweaty chest to absorb his "amniotic" fluid. *Couvade* can be seen as diminishing the disparities between male as masculine and female as feminine: teaching the male to be nourishing and compassionate.[33] The very terms masculine and feminine impede the cultivation of "a more consentient archetypal form."[34] John Muir:

> Drinking this champagne water is pure pleasure, so is breathing the living air, and every movement of limbs is pleasure, while the whole body seems to feel beauty when exposed to it as it feels the campfire and sunshine, entering not by the eyes alone . . . One's body then seems homogeneous throughout, sound as a crystal.[35]

Muir's ability to relate empathically to the Earth, not just with a squinting set of exteroceptors, rendered him sound. He did not habitually objectify his body, dividing its surface into the five "outer organs," distancing himself from the Earth.

Vital individuality requires visceral responsiveness of the whole self to the whole Earth, to here and beyond, now and not now, the inclusive female aspect. John Keats writes in "I Stood Tiptoe on a Little Hill,"

> *Linger awhile upon some bending planks*
> *That lean against a streamlet's rushy banks . . .*
> *How silent comes the water around that bend;*
> *Not the minutest whisper does it send*
> *To the overhanging sallows . . .*

The water comes silently from the beyond. Keat's responsiveness might be called female, and he no less a man. Or a woman's sharply demarcated individuality might be male, she no less a woman. Archetypal myths and metaphors are powerfully revealing: true as art because disclosing multiple veins of potentiality. Mother Nature, then, is true in the ancient senses: touching archaic memories into new life, she shows things' interfusion and their distinctive capacities for development, both.

The sense of Mother Nature or Earth Goddess actually absorbed in our bodies is much more powerful than what we commonly construe today as metaphor. Undermined is the childish sense that we already grasp literal meanings, and that the arts (and religion) are distractions or amusements, that which keeps us boxed up in patriarchal alienation and fury. We are huge, large-headed, infantile beings thrashing and clutching, but no longer appropriately so—out and about in Earth-Mother now, separated from mother, more dangerous than actual children, and needy and angry.

Standing open to the darkness of one's earthen body leaves room for the unknown. If from the beginning each leaves room for the unknown in the other, each lets others be—within their limitations, of course, whatever these turn out to be.[36] If so, there is no call for frenzy or aggression. The root of evil—of addictions, for example—is the fear of impotence or annihilation.

After reading of Dewey's long encounter with Alexander, I investigated such a regimen for myself. I was unhappy with my life. I had not smoked for many years, and was not addicted to alcohol or hallucinogenic drugs. But I remained a workaholic. I had put far too many eggs in the basket of professional success. Only after achieving some of it did I realize the mistake. My life was lived at a dead run. Many times others had needed my care, but that had made no difference. I ran past them, or through them.

William James disclosed that there is something wild and heroic at the heart of us, but I had not grasped how the need for heroic accomplishment in my own life had constricted it. I had become addicted to a perverted mythic ecstasy that drove my body as if it were a machine, throttling and strangling me and hurting others—a patriarchal addiction supplying endless rationalizations for its existence, garbed as it was in brilliant heroic insignia.

I had heard of a variant of Alexander's approach, developed by the Israeli polymath Moishe Feldenkrais, which required recumbent exploratory exercises. The aim was to be free of automatisms and malcoordinations acquired in first learning to stand up; to return to a state which, if not completely unjelled, allows us to start over again, to undo makeshift satisfactions of ambulatory and postural needs that encage us adults. This sounded right and I signed up with a Feldenkrais practitioner.

The first lessons appeared pointless. I made microscopic movements with my lower back by rotating, arching, and lowering it to describe a small circle on the floor. The goal, the woman instructor said, was not speed or accuracy but simply awareness of what I was doing. At times a bit of levity—"Arch your back just enough to let a furry little creature scurry under it." She tirelessly questioned, lesson after lesson, "Are you breathing?," "Are you breathing?" In consternation I usually had to answer that I was not.

So that was the point, simply to become aware of my body in its incredible closeness and intimacy. I found that faced with any difficulty I automatically and mindlessly held my breath, either momentarily with a small problem or for surprisingly long periods with something obviously frightening or repugnant. I was saying No to it at the primal level by not allowing it, its atmosphere, its presence, into my body. I turned myself addictively into a tight, frightened fist. I failed to breathe because I feared to incorporate the presence of others.

James's and Dewey's ideas about body-self and body-mind were finally confirmed for me. I found that by simply allowing myself to breathe, the repugnance and fearsomeness of things diminished. As my disgust and fear decreased, so did the thing's disgustingness and fearsomeness.

The next exercise involved making microscopic movements in my viscera. Though I was breathing a little more regularly and effortlessly in some situations, I discovered unexpected tensions in the viscera. Deliberately opening myself to the surround, dilating attention beyond objects of immediate concern or use, my viscera also opened and relaxed for periods of time. Then I would feel in tune with the humming world, harmonized, birthed by it. Without these exploratory movements of the body, I would never have become aware of constrictions that I—as a deeply socialized and technologized being—had created.

Certain spastic clots slowly dissolved, and black holes of paralyzed breathing were less common. At times I used the image of a slowly cycling clockhand to aid voluntary-involuntary breathing. A living bridging, a webbing. I remembered—re-membered.[37]

The conditions of consciousness typically lie beyond consciousness. The body-self's self-awareness is egregiously limited. Only when a certain habitual movement or posture of the body is opposed gently by a countermovement can the habit be exposed to an expanded consciousness. But before its exposure, the person does not know what is to be opposed or turned. So how to begin?

Typically, reflective or calculative consciousness is powerless, and some step in the dark by body-self is needed to feed back the body-self revealingly into itself—truth as primal, sensuous showing. Another may have to take us by the hand so that body-self's constricting and entrancing habits are sensitively opposed and thereby exposed. That a difficulty cannot be directly opposed and defeated through one's own heroic efforts is difficult or impossible for the patriarchal mentality to fathom.

Feldenkrais estimated that 80 percent of behavior is determined by the habitual movements of the eyes. So if this behavior is to be exposed and perhaps changed, the eyes' automatisms must be gently opposed. The eyes typically turn the way the body is turned, so his exercises turn them the opposite way. When the eyes' constricting habits are lovingly turned, they are exposed, and consciousness dilates—just as when the holding of breath is eased and opposed, it is exposed, and the viscera can relax. When we are relaxed and unthreatening, the threat to ourselves diminishes. When we refuse to dominate, the fear of being dominated dwindles. We can oppose without trying to crush.

Through opening to the world, the world opens to us. As if one were a wise child and the universe a mother, there is interdependency. The Mother turns us gently when we are bent on rushing over the edge. The test of an art of life and its truth is whether it allows circular power to return fluently into itself through ourselves. When it does, the spell of addictive behavior is broken.

Harmonizing and Phrasing Voluntary and Involuntary
Andrew Weil, M.D., illuminates the involuntary nervous system, and the need to wait upon it if we would weave it into voluntary life and our full sense of self. He respects the inclusive and maternal, the matrix:

> Breathing is the best example of a function with dual innervation: our respiration can be totally conscious or totally unconscious. In one case, the voluntary nerves run things; in the other case, the autonomic system carries the impulses.[38]

Weil undercuts patriarchal and dualist thinking that sunders mind, consciousness, and the voluntary from body, unconsciousness, and the involuntary. It is only the consciousness involved in conventional Western science and industry—ego-consciousness—that cannot directly control the autonomic functions. Consciousness relaxed—trancelike hypnotic or yogic consciousness—can influence the autonomic system, prompting the voluntary and involuntary to dance together.

> Respiration is the one function in which the two motor pathways are in perfect potential balance. The theory behind yogic and other systems of disciplined breathing is that regular rhythms produced by the voluntary pathway will eventually be picked up by the involuntary pathway, and that once this sort of correspondence is established . . . it spreads naturally to the other involuntary functions of heart rate, circulation, and so forth.

The voluntary and involuntary, mind and body, are but two aspects of one energic reality. And, no surprise, it is the viscera that hold the

secret of integrating body-self into a flowing and momentous way of
being.

> Becoming aware of internal functions means paying attention to sensa-
> tions we ignore in our ordinary waking state. As defined by neuroanato-
> mists, the autonomic nervous system is purely motor . . . (that is, it carries
> impulses away from the brain . . .). But it is well known that sensory
> nerves travel with this system, carrying information from the internal
> organs to the brain. These visceral afferent nerves . . . are among the least
> well understood components of the nervous system . . . The experiential
> correlate of this lack of understanding is our own inability to feel what
> goes on in our bodies except in the vaguest ways. . . . These visceral
> sensations are diffuse (that is, poorly localized) . . . Yet Yogis say that with
> practiced concentration, one can greatly sharpen one's perception of
> these signals and that success in this practice leads automatically to
> greater autonomic control.

This concentration is a spontaneous doing-and-allowing. For without relaxing and allowing the vast world to excite resonances in the
viscera, we cannot know where each of us belongs within the world. We
are alienated within ourselves.

*We were walking in England's Lake Country. Noticing a horse in a pasture we
stopped. We were fortunate to see it at all, for it stood stock still under a tree,
looking out at us from under boughs. All alone, the animal looked forsaken.
Standing at the rail fence, my wife and I tried various appeals to lure the creature toward us. For several minutes we tried, the horse keeping its frozen stance,
baneful and pitiful. Our daughter, about eighteen at the time, made a few little
sounds in her throat and the horse came over.*

*When she was little—about the time we had Buppie—she collected and cared
for small animals, rabbits and abandoned cats and kittens. Under her care, one
apparently undistinguished rabbit was so handsomely groomed and nourished
that the judge at the local fair created a special category for an award: General
Rabbit. I referred to it as such at every opportunity.*

*The cats roamed the house and formed our daughter's most intimate family,
which amused me. But I did not relish them roaming our bedroom as we slept.
So each night, however tired, I tracked them down and ushered them, a bit
roughly at times, into the furnace room.*

*One of her cats—Charcoal—became sick and emaciated. She dragged her
nearly paralyzed hindquarters. Clearly she would die soon. One evening, late,
while all others slept, she looked up at me and made a sound. I found myself*

kneeling beside her and holding my head against hers. Astonished and elated, I
said good-bye to her.
  Now that our daughter has died, I can only open to her presence each day,
and try to retain what she brought us.
  You stirred into celebration everything you touched.

The Sacramental and the Sacred

Becoming aware of internal functions can be an experience of the sacred. Sacraments without the Female aspect are eccentric. Conger
Beasley's experience suggests a deeper sacramental life.[39] He recounts
accompanying an official of the Alaska Department of Fish and Game
on an expedition on the Bering Sea. The goal was to shoot four seals
so biologists could analyze blood and tissue samples for toxics, trace
minerals, and parasites.

  Revolted by the experience, Beasley clutches for some redeeming
qualities in it. After a seal is shot, its blood boils up around it in the icy
water. But the redeeming feature is there: for the first time Beasley
realizes viscerally his consanguinity with seals. As they open up the
seal's abdomen and extirpate its vital organs, Beaseley notes,

> I developed an identification with the animal that carried far beyond
> mere scientific inquiry . . . the abdomen of an adult harbor seal is approx-
> imately the size of an adult human male's. Each time I reached into the
> tangled viscera, I felt as if I were reaching for something deep inside
> myself. As I picked through the sticky folds of the seal's heart collecting
> worms, I felt my own heart sputter and knock.

Beasley realizes—viscerally—that "the physical body contains func-
tional properties, the proper acknowledgment of which transforms
them into a fresh order of sacraments." Coiled intestines intertwine
with coiled intestines of all animate things. In the recoiling intake of
air, in the involuntary gasp of awe induced in our bodies, we pay tribute
to the wilderness *mana* energies shared with all animals. We let them
into us. The sacrament is the involuntary acknowledgment of our
kinship and preciousness that resonates, nevertheless, through our voluntary consciousness and career. It is sacrifice of ego: the acknowledgment of all that we do not know and cannot control, and upon which
we depend. It names and appropriates the sacred.

  The lack of ritual means of coordinating the visceral center of our
being with the regenerative cycles of kindred beings in wilderness Na-

ture is impoverishing. No wonder that narcotics addictions run rampant in a society that seems to have everything.

The powers of the taboo, holy, or sacred increase in direct proportion to our efforts to addictively control, ridicule, or ignore them. But they emerge grotesquely, wild in the worst sense. Everything may be "under control," but the cocaine is in the executive's desk drawer.

The Spiraling Matrix: Taoism

Gary Snyder writes of

> . . . the muse point of view . . . that became covered over: It's in Taoism, and within emphasis on the female . . . the spirit of the valley, the *yin*. Taoism being, following Dr. Joseph Needham's assessment of it in *Science and Civilization in China*, the largest single coherent chunk of matrilineal descent, mother consciousness-oriented, neolithic culture that went through the, so to speak, sound barrier of civilization in the Iron Age and came out the other side halfway intact.[40]

Essential to Taoism is the ritualized meditation in movement called T'ai-chi-Ch'uan. *Chi* means energy, and the flowing dancelike positions of "the Form" reconnect human animals within the regenerative energic matrix and hence within themselves. Anxious ego tendance is sacrificed and the world experienced is allowed to flow through us. Tribute is paid viscerally to the wilderness *mana* energies shared with all things. Gratification happens. Deep and lasting gratification cannot be forced.

In Tai chi's evolving configurations of body-self, energy is rooted through the feet, developed through the legs, directed by the viscera, and extended by the arms and hands. We feel breath coming up from the Earth, through the body, out the top of the head, and, exhaling, flowing back down through the body and back into the Earth, and on and on. This is the way it feels. It is action as allowing: being so rhythmical, receptive, poised, and open—yin-like, female—that the effects of movements are fed back into the system and form the production of the next movement, and so on. The process is seamlessly whole.

Centered in the viscera as we move, and turning around the axis of the backbone, we are realigned with the axis of the world, *axis mundi*. Once we are flowing in the dancelike attitudes of the Form, the urge to control by might or main or shrewdness is seen to be stupid. T'ai chi epitomizes positive freedom: the ability to do the good thing, which is to be in flowing harmony with the rest of the world.

Mary Midgley:

Conscience is not a colonial governor imposing alien norms. . . . It is our
nature itself becoming aware of its own underlying pattern. . . . It is not
free to make up the rules of the value game.[41]

The ability to harmonize emanates out of the *Tan tien*, in the viscera
two inches below the navel. Here is ground zero, a flowing, ever-level
center, sinuous and spontaneous, keeping to its own centeredness
within the plurality and interdependence of all things. As the body experiences itself poised in the initial tremulous stillness of the first position of the Form, it is very like a cobra rising erect from its coils, vibrant
with potentiality. It is the primal wilderness body, but now caught up
in the disciplines of later civilization, become civilization's intimate
other. In awed silence, the regenerative presences of things are held
coiled within ourselves. In T'ai chi art and religion are one, dance and
reverence one, the body-self one.

This way of life comes to us from ancient times when shamanic healers enacted living things with a verisimilitude deriving from human
bodies' ceremonial incorporation of other living animal bodies.[42] To be
poised is to be caught up in the pulsing womb of the world, space a
matter of time, time a matter of space, bodies interweaving with each
other.

Today, all over China, numerous people of all ages, castes, genders,
temperaments can be seen doing their T'chi, meditations in movement,
in parks, fields, or valleys.

Female Initiation Rites at Puberty: Can We Learn Anything?

Reclaiming a fitting form of ritual life is badly needed. Coordinating
our heads, bristling with ideas, with the body's empathic trunk and the
great emotions—what could be more important or difficult? Either this
happens or individual and corporate egos, caught in their mind games
and control manias, will unhinge Earth's life.

Let us look at some deeply rooted adolescent female initiation rites.
Some threads run through them all. These derive in one way or other
from the momentous transition into menstruation, the sign that the
young woman can bear children (*ritu* in Sanskrit, from which our *ritual*
is derived, means *menses* [literally months in Latin] or menstruation—so
fundamental is menstruation to forming the idea of ritualistic life[43]).
The woman's cycle of fertility is part and parcel of the moon's, and
the moon's cycles constitute the lunar year, the earth's yearly cycle of

regenerativity: birth, in spring, then growth, harvest (or with animals, rutting), decline back into the earth, death or dormancy, rebirth. In many traditional societies women living closely together with one another and the surrounding world—ingesting odors and pheromones— apparently menstruated together, and at the dark of the moon.

As nobody can command the phases of the moon, so nobody can command the onset of menses. It was understood that one must simply wait expectently, reverently, sacramentally.[44] So far this lies from our technologized world! But the relevance to us is plain when the first routinized surface of appearance is peeled off. For though our chemical interventions can control fertility to some extent, and even shift somewhat the onset of menses, the fact of fertility still exists, and we either find it meaningful and rejuvenating or distracting and wearying. Technology by itself cannot help us here.

Note a female puberty rite of a hunter-gatherer society extant in the Northwest Amazon: "Festa das Mocas Novas: The Cosmic Tour."[45] The community derives its food largely from fishing and uses an herb, timbo, that "changes the surface tension of the water and causes fish to die of suffocation."[46] If they use too much of it, however, they kill too many fish and destroy their fishery and their lives. Life for them is a matter of fine balances.

Female fertility is, of course, a blessing, for without it the group would soon cease to be. But it is also a threat, because if too many children are born, the group will outstrip its food supply. Faced with this balancing challenge, they have developed a rite astoundingly complex and effective:

> On the one hand, the initiand is celebrated, made eligible for sexuality and marriage, and encouraged to produce new life through the example of the new mother in the myth of the origin of masks who escaped the wrath of the Noo by virtue of having just given birth. On the other hand, there are mythic exemplars who provide . . . quite a different model. No children are reported for Ariana and no lovers for the virgin Mother of Timbo. Furthermore, the [initiand] is required to be a virgin at the time of initiation, and within the ceremony conscious steps are taken to inhibit her future fertility. She is given an upward flowing bath by shamans, which reverses the direction of birth, and she concludes the rite with a contraceptive bath of timbo . . . . By renouncing for a time the conception of children, she benefits the world of nature, permitting the fish population to thrive. But she also makes it possible for people to avail them-

selves of this increased supply of fish: she becomes the Mother of Timbo, the most efficient means of fishing.[47]

A rite simultaneously of conception and contraception! The members of the group are sound and coherent individuals, willingly conforming to the regenerative whole of which they are parts. Very probably they experience their urges as through and through their own. They are not fractured and befuddled by the willfull will to control.

It is not clear how a mind-set similar to theirs can be inculcated in us today, enclosed as we are in a juggernaut of production which may have already generated irreversible degradation of the earth and our bodily selves. Limited as we are, we still might come to see our limitations, and move into new horizons of receptivity and awareness. We are addicted to prosaic meaning and truth, and to tunnel vision linear thinking, only if this is all we can imagine.

A Partial Account Pretending to be Whole
Western science and technology, brilliant in their own ways, totalize their claims, like a transcendent male god who surveys the world, confident that what lies beyond horizons, the unknown, will sooner or later yield to knowledge and probably control.

But I sense in myself, and in others, a dim awareness of the unknown and uncontrollable. In moments of vulnerability we resonate in dread to what lies beyond horizons. Nature is not merely raw material to be exploited for short term gain by North Atlantic technological genius. No matter how adaptive and productive technological-commercial activity initially appears, its "adaptive" numbness blots out awareness of the finitude and mortality of all living things. The Great Mother will reclaim each of us.

The "me and not me" experience is addictive trance. Since individual and corporate individual mirror each other, we should expect to find "we and not we," "us and not us" experiences disorienting the group. As a self-deceiving individual tips to the "not me" side and evades responsibility and guilt, so the self-deceiving corporate individual tips to the "not we," "not us" side, and does likewise.[48]

This happens—can we doubt it? For a prime example, international corporations typically work in a trance of constricted vision, unwilling and unable to see the broader consequences of what they do to earth and earthlings. Most are caught up irresponsibly in immediate profits, fixes that must be addictively repeated. If it is hard for individuals to break spells of addiction, it seems at least that it is harder for corporate ones.

We are challenged even to imagine what has not happened yet, and may never: global or cosmic democracy, in which nearly everything is tolerated except intolerance itself—in its most insidious form, intolerance of the regenerative cycles of Nature, our own bodies included. Uprooted from common source in myth and ritual—even from the myth of democracy—we tend to limit the democratic life to certain political forms.

Sudden changes in group structure and consciousness—the collapse of the Soviet Union is a fine example—afford some hope. As individuals sometimes swerve dramatically and unpredictably, so corporate ones may likewise. Attitudes and reactions are sometimes greatly contagious, resonances often surprising.[49] It is not inconceivable that fashions develop suddenly in corporate individuals for celebrating conception and contraception simultaneously, so to speak. These would be de facto rites as sophisticated and fitting to our situation as the hunter-gatherers' are to theirs.

Again, as Muir said, "everything is hitched to everything else." Who can say where consequences will end when an individual changes his or her own life?

> I will not let you go until you bless me.
> —JACOB TO THE ANGEL, GENESIS 32:36.

Notes

1. But patriarchal bias is not limited to males. Women brought up within it typically absorb these same values: they place a greater value on males than on females, offering distinctive services to males. Concerning the different ways that male infants are treated by mothers, see Nancy Chodorow, *The Reproduction of Mothering.* Concerning the wrenching effect on the identities of women talented in science, but initiated into the ruling paradigm of "Active Male Mind" and "Passive Female Nature," see Bat-Ami Bar On commenting on E. F. Keller in Alcoff and Potter, *Feminist Epistemologies,* 91.

2. It is a cultural fact of long-standing that the early primary care-giver has been the mother, or a female stand-in for her. This does not mean that males cannot play the "motherly" role, but the motherly or Mother archetype, as it has evolved to this point, still exerts great influence on everyone's attitudes. It is impossible to predict how life will change if this ceases to be the case.

3. Joseph Pearce, *Magical Child*, 226.
4. 226–7. Also see Luce Irigaray's trenchant analysis of male resentment toward woman as a reaction formation to nullify the desire to move back inside her: "Veiling his nostalgia in contempt. And vomiting up that first nurse whose milk and blood he has drunk" (*Marine Lover*, 26). Again, directing her critique explicitly at Nietzsche, "You are now immersed and reenveloped in something that erases all boundaries. Carried away by the waves . . . Tragic castaway in . . . turmoil (36).
5. "Matristic" is Marija Gimbutus's term. It is preferable to "matriarchal," for that suggests the patriarchal hierarchy. Women simply replace men in the slots.
6. For a brilliant and learned account of the rise of patriarchy, see Thorkild Jacobsen, *The Treasures of Darkness*.
7. For example, it was customary in classical Athens to denigrate the "passive" member of a homosexual pair, but not the "active." See Arthur Evans, *The God of Ecstasy*, New York, 1988, 94–96, which gives examples of the stigma of being penetrated—one that helped allay male identity anxiety. "Active" members of homosexual pairs were allowed to testify in court—were presumed to have authority—but not "passive" ones.
8. See, for example, Carol Gilligan, *In a Different Voice*, and Alison Jaggar, *Feminist Politics and Human Nature*. Of course, some women feel aversion to their capacity to give birth. As one put it, "It's like shitting a watermelon. Who needs it?" But, I think, the capacity to bear children does incline toward distinctive behavioral tendencies of a general sort in the long run in large populations of women—and this is probably true across cultures. Consult Charles Peirce's notion of universals.
9. Erik Erikson, following Freud, thinks that women's womb is an "inner productive space" that men do not have. And women tend to fear being left empty or deprived of treasures or unfulfilled or drying up. (In Sered, *Priestess, Mother, Sacred Sister*, 286 fn. 1). This framing of physiognomic differences entailing "psychical" ones seems to reflect male bias. Even women never pregnant can feel full of womb blood every month and full of potential. Moreover, in the crone or wise woman tradition, after menopause the woman keeps her wise blood inside (Donna Wilshire, *Virgin, Mother, Crone*, 166). If Freud and Erikson were right, and if addiction entails unsatisfied needs—a lack—shouldn't we expect to find more women addicted than men? That does not seem to be.
10. The distinctive ways in which women feel their own bodies from within them, kinesthetically, is described by Luce Irigaray in *The Sex that is not one*.
11. Magnetic resonance images of men's and women's brains show differences in their decoding of words (*Nature*, Februrary, 16, 1995, and *New York Times*, same date, front page). All nineteen men in the study used "the speech area" in the left side of the brain (note the assumption that what is true of the male is normative for the species). But eleven of nineteen women used both this area and a comparable area on the right side, suggesting at least that many women tend to be more integrative when using words (which we might have intuited anyway?). Also see Ruben Gur (*Sci-*

ence, January 27, 1995): Injecting radioactive glucose into men and women, and using a machine to measure brain activity similar to a CAT Scan, Gur discovered that men had higher activity in the limbic system, a part of the brain used in emotional processing that remains from the era when reptiles flourished, and our prehuman structures were taking shape. Reptiles do not think before lashing out. That most wars are started by men may not be a "mere cultural accident."

12. I try to describe the body in its immediacy, undivorced from our sense of self. This is sometimes referred to—not perfectly happily—as body-image (See Paul Schilder, *The Image and Appearance of the Human Body*). Elizabeth Grosz (*Volatile Bodies*) exposes how most male writers simply assume that the male body is the norm. She gets into the necessary specifics: "In discussing the libidinal structure of the body image, Schilder states that the unification of the body-image and the cohesion of our self-identities is dependent on the attainment of a stable, genital form of sexuality. While this seems true for men and male sexuality—insofar as genital, phallic sexuality hierarchically subordinates but does not eliminate the pregenital drives—it is not even clear what this would mean for women and female sexuality—insofar as female sexuality is already *genitally* multilocational, plural, ambiguous, polymorphous, and not clearly able to subordinate the earlier stages" (83). (Also see Sered, *Priestess, Mother, Sacred Sister,* 138.) This has great implications not only for how men and women are to be known, but in how we *know* the world—what it can mean for us.

13. This is perfectly clear only in his fugitive poetry. See Jo Ann Boydston, *The Poems of John Dewey,* particularly "Two Births," and the untitled, "Now night, mother soul, broods the weary hours" (first line).

14. W. T. Greenough and J. E. Black, "Induction of Brain Structure by Experience . . ." in *Developmental and Behavioral Neuroscience,* 24 [1992]: 155–299. Also I. J. Weiler, et al., "Morphogenesis in Memory Formation: Synaptic and Cellular Mechanisms," in *Behavioral Brain Research,* 66 [1995]: 1–6.

15. Dewey, *Experience and Nature,* chap. 2, "The Precarious and the Stable."

16. Carol Bigwood (Sources), her critique of Judith Butler. See particularly chaps. 1 and 2, "Is 'Woman' Dead?" and "Renaturalizing Gender (with the help of Merleau-Ponty)."

17. *The Roots of Power,* 84–94.

18. 82.

19. Sheets-Johnstone offers examples of primate behaviors embedded in current human life that are less modified by cultural constraints: the dominant individual's fixed stare that intimidates subordinates, or any individual's creation of a commotion to attract attention to itself for whatever reason. Also see primatoloist Frans de Waal's sensible conclusion that "culture and education mold gender roles by acting on genetic predispositions" (*Good Natured,* 121). And given that every mammal that has survived in 200 million years of evolution has received maternal care, it is no wonder that human females tend "to value intimacy, care, and interpersonal commitment" (123). Also "Studies consistently demonstrate higher levels of emotional contagion in female than in male infants" (121).

20. See Chris Knight, *Blood Relations*.

21. See Donna Haraway, *The Cyborg*.

22. Iris M. Young, *Throwing Like a Girl*.

23. See Robert Wright, *The Moral Animal*, 33 ff, for a summary statement of "evolutionary psychology" on this issue.

24. As Edward Casey has pointed out, the word *matrix* undercuts the distinction between matter—supposedly objective—and form—supposedly imposed by subject or mind. (*Remembering*, 294–5.)

25. But, emphatically, this is not addicts' experience of bodily urges being "both me and not me." Apparently, when the woman experiences the separateness of the fetal body "it is as it should be." Carol Gilligan (*In a Different Voice*) offers extensive research into the ways girls resolve disputes and address issues in general. It tends to be less oppositional and exclusive, less a zero-sum game, than it tends to be with boys. Some will say this is merely cultural conditioning; the view I am developing, however, rejects all exclusivistic "either/ors." The matter is not either biological or cultural. I discard the residue of the Cartesian program to break down things into the "ultimate simples" out of which the world is built—in the pervasive Cartesian framework, mental or physical simples. Human development occurs in the murkiest intersection of biology and culture (recent research into how cultural influences mold the development of the brain bears this out). Attempting to break down a problem into calculable factors reflects an unwillingness to confront the incalculable level of the World-whole. In fact, we cannot predict what will happen as archetypes shift.

26. Sheets-Johnstone writes, "To denigrate the involuntary is to miss the miraculousness of Nature" (333).

27. For example, see Neihardt, Sources, *Black Elk Speaks*, in which Black Elk refers to the community of beings. See also Catherine Keller, Sources.

28. Are men, *on average*, less suited today to play the motherly role than are women, *on average*. I believe so.

29. But see de Waal's multi-aspectual analysis of the leader's cooperation and caring in both maintaining his position *and* in gradually loosening the hierarchical structure (144).

30. See Glen A. Mazis, *Trickster, Magician, Grieving Man*, chap. 4, "The Masculine Fear of Words of Feeling."

31. *Priestess, Mother, Sacred Sister*, 9, 114 ff, 195 ff.

32. *The Life Cycle Completed*, 9.

33. Nor Hall, *Broodmales*.

34. Sheets-Johnstone, *Roots of Power*, 331. Naomi Goldenberg decries Jung's static opposition, masculine/feminine. "We must redefine archetype and experiment with new attitudes toward myth and iconography." (*Changing of the Gods*, 62–64). I agree.

35. *My First Summer in the Sierra*, condensed in *Gentle Wilderness*, 55.

36. Luce Irigaray, *I Love To You*, 117.

37. For a more recent variant on these methods, see Joseph Heller and William Henkin, *Bodywise*.

38. *The Natural Mind*. The three quotations are taken from 160 and following.

39. "In Animals We Find Ourselves," *Orion*, Summer, 1990. That quite a few need a clue to a deeper sacramental life was evident in a letter to the *New York Times* (November 13, 1992) written by Yale professor, Orlando Patterson, in response to the coverage of the Clarence Thomas/Anita Hill confrontation before the U.S. Senate Judiciary Committee. He said that women like Hill were neo-puritans in the grip of a "reactionary sacralization of their bodies." If they want political power, he went on, they will have to give this up. But at what cost to their and our ecstatic life?

40. *The Old Ways*, 38. I once heard John Muir quoted to this effect: "When the White man's steel axe rang out on a startled world . . .".

41. *Beast and Man*, 274.

42. Concerning still surviving abilities of this mimetic sort among Australian Aboriginals, see Bruce Chatwin, *The Song Lines*.

43. I owe this etymology to Elinor Gadon, *Once and Future Goddess*, 2.

44. See Donna Wilshire, *Virgin, Mother, Crone*, especially the sections, "Virgin Consciousness," "Mother Consciousness," "Crone Consciousness," in which ancient woman-centered rites of passage are elaborated. This account includes her performance scripts (also available enacted on audio cassette).

45. Bruce Lincoln, *Emerging from the Chrysalis*, 50 ff.

46. 67.

47. 69.

48. So perfect an example of corporate self-deception is the group-body of Nazi medical doctors that we may forget that this sort of corporate phenomenon is very common (and resultant genocides are not uncommon). But Robert Jay Lifton's analysis of the Nazi corporate body (*Volkskörper*) exposes brilliantly the general pattern of corporate circumpressure. In the Nazi case, many of the very people dedicated to healing were prompted to rationalize medical experiments (tortures) and murder. Corporate and individual self-deception excite and reinforce each other. Any imagined or perceived threat to the health of the corporate body (e.g., Jews as "germs") rationalizes genocide. *The Nazi Doctors*, 29, 36, 45–6, 62, 84, 184, 433.

49. Appraise carefully Rupert Sheldrake's idea of morphic resonance, in Renée Weber, *Dialogues with Scientists and Sages*, 71 ff. Also see John Perkins, *Shape Shifting*. Perkins applied shapeshifting techniques to his career as a management consultant and president of a U.S. energy company, founding an organization that inspires executives to clean up pollution, reshape corporate goals, and form Earth-honoring partnerships with indigenous cultures. The reader may blink. Is such a thing happening? It is, apparently. Most intellectuals, established as such in the society, wince at such an account. But history is replete with examples of established intellectuals being out of touch with the most creative members of their societies.

Chapter Twelve

Technology As Ecstasy:

How Do We Deal with It?

Although technology is ages old, and has developed through thousands of years, until the end of the eighteenth century it remained a part of handicraft. Hence it remained basically unchanged and part of man's daily life within his natural environment. Then, during the last 150 years, technology made an incision deeper than all the events of world history over the past thousands of years, as deep perhaps as that caused by the discovery of tools and fire. Technology has become an independent giant. It grows and advances. It brings about a unified and planned exploitation of the globe . . . which is financially profitable. Trapped in the spell of technology, men seem no longer capable of controlling what originated as their own works. . . . The restless march of technological change on a gigantic scale makes us stagger between ecstasy and bewilderment, between the most fabulous power and the most elementary help-lessness.[1]

—KARL JASPERS, *THE IDEA OF THE UNIVERSITY*

Technology is as much a part of us as are our hands, legs, eyes, and ears; and minds and selves are not separate from bodies. To extend **spiritual body-selves humans have always used tools. Some other animals use them to some extent, but they play a massive and distinctive role in human life. Since it is essential that we stand out from our bodies into what lies beyond us and be absorbed ecstatically in possibility and actuality, and since tools are essential to this, technology is itself ecstatic.

But technology is a threat because its ecstatic reach absolutizes itself, peripheralizing other ecstatic capacities. It cannot adequately substitute for them; it even tends to erode visceral and empathic bonding with many of the things in which it involves us. Either the brightness of new technology eclipses other ecstatic modes of being, or the stupefaction of routine technology dulls the self. Technology—new or routine—is an avalanche that buries the quiet and sustaining ecstasies of the sacramental life. These were the ancient ones in which humans delighted in simply being-with everything else. These were the ritualized ecstasies that bound people into the Whole and rendered their lives momentous and coherent.

In the garb of marvels—limitless atomic power, spy satellites, e-mail and television—technology conceals sacramental openings onto the world, and conceals the concealing. Claiming to give access to the World-whole and to place us meaningfully within it, technology can only deceive us into thinking that the principle of value is satisfied: the greatest differentiation and individuation within the greatest coherence and unity.

In moments of technologically facilitated excitement we may believe this principle has been satisfied, but the next moment sees the belief collapse into what we half-suspected all along: the absence of the sacred's mixture of excitement, expansion, fear, restraint—the absence of awe. As Jaspers wrote, "We stagger between ecstasy and bewilderment, between the most fabulous power and the most abject helplessness," and become easy prey for demagogues who—decisive!—claim to lead us out of confusion.

Robert Bourassa, former premier, Quebec:

> Quebec is a vast hydroelectric plant in the bud, and every day millions of potential kilowatt-hours flow downhill and out to sea. What a waste![1]

Technological ecstasies resemble a dynamo's eccentric flywheel that will tear itself and its housing to pieces with a few more RPM. There is widespread addiction, from workaholism leading to burnout, to the grimmest drug use.

Certain technologies, however, might abet a larger ecstatic and regenerative life instead of being addictive substitutes. They might enhance that wisdom of the body in which meaning is made, balance perpetually recreated. But recall Vincent Dole's warning mentioned in chapter 1: we sometimes throw ourselves out of balance in order to experience the excitement of trying to regain it. How to deal with danger? Try to suppress it and live dulled lives? Or blindly indulge in it, shootingly wildly for satisfaction?

The Arrogance of Modern Technology

To travel in a speeding machine faster than any animal can attain on its own—at times this is an ecstatic, regenerating experience. Or to hang glide, to fashion wings and soar in thermals, mimicking eagles or condors, perhaps simulating their telescopic eyes with binoculars attached to our heads. Or to free fall from planes, doing ring dances at 10,000 feet.

Nevertheless, modern technology in its ubiquitous routine use tends to inculcate habits of mind that grossly exaggerate abilities to enhance our lives through direct control. Plankton, other plants, animals, insects deeply affect the rest of Nature. Changes they make are profound, but typically slow, giving Earth time to adjust itself and reach new equilibrium. We abruptly initiate radical changes. And as soon as the technologically altered situation becomes routine, it tends to be regarded as normal, no matter how great the disruption in the balance of things. Despite being creatures who can feel and imagine beyond the immediate horizon, and construct and reconstruct accordingly, we become walled up in unseen limits. Unable to assess most of the consequences of our alterations, and confined in spaces whose limits are not seen, people plunge like bulls in china shops. Technology, begun as handicraft, gets out of hand.

Primal Tools/Modern Tools

As one of her startling discoveries from the earlier Neolithic, Marija Gimbutas describes axes. Some seem too small for utilitarian purposes, others too large. We would call them symbolical.

But the very distinction between symbolical and utilitarian is probably a projection on our part. For the shape of these axes is triangular, symbolizing apparently the female pubic region, that numinous center for creative and recreative power within Paleolithic and early Neolithic experience, the mark and presence of the Female. That is, some axes were used for cutting, but for those people cutting was participating in the numinous creative and recreative energies of the universe, Circular power. For us, cutting is not, and we should not be surprised that it runs away with itself.[2]

The great residue of modern technology become routine confines itself within habits of instrumental and calculative reasoning. If A is wanted, technological means B will achieve it. But technology by itself cannot determine that A ought to be produced for its own sake. Even though technology is essential for producing A, it cannot by itself demonstrate A's intrinsic value. Only if A is itself a means to something else can technology assess its value as effective and efficient means to secure that.

But sooner or later an intrinsically valuable "something else" must be found, or technology will have no target. It will simply be the means for more means for more means; power for more power for more power . . . without limit and measure, boring blindly and addictedly into the world.

Technology's tools extend in some ways the primal, ecstatic body. We neither neatly objectify it as an object that annexes another object, a tool, to itself. Nor do we neatly subjectify it and address tools in a fully deliberate and self-conscious way, "Now I, at this point in my career, take up this tool in order to accomplish A, and will be held accountable by criteria x, y, z." Typically, we are possessed by tools ecstatically, acting with them in a kind of trance: sometimes that of routine and boredom, at other times of angelic or demonic possession. What of the whirring and roaring of saws in great forests and jungles? Caught, typically, in a kind of continuous patriarchal daydream, lumbermen cannot imagine the possibility of awakening. They are not in possession of themselves, are addictively infested.

To live in a world out of control is to feel stress with no way of relieving it. To dull the pain we "adapt" in addiction: something is satisfied, ephemerally and marginally. The society's addictions and the individual's fuel each other.

Modern Technology's Blindness to Its Own Mythic Assumptions

Though technology by itself cannot perceive and certify intrinsic value, we, its operatives, unwittingly entertain numinous presences—dazzling archetypal presences, beings we implicitly regard as intrinsically and unconditionally valuable. Science and technology, in addition to their demonstrable powers, also play this mythic role tacitly. What goblins of excitement and acquisitiveness possess those who clear-cut ancient forests?

The rush and trance of avowedly secular technology reenacts mindlessly powerful religious motives, myths, deracinated sacraments. It is

religion as technique that acts out the myth of progress: the exploitation of physical Nature through purely instrumental reasoning. Spe-
cifically, mass production is the ritual embodiment of the myth of
progress. One more item, one more, one more . . . the dogma that ever
accelerating growth (of a technologically appraisable kind) is good.
Mass production's repetitions are demonically tinged—the corporate
body's behavioral addiction.[3]

John Dewey writes:

> We live in a world in which there is an immense amount of organization,
> but it is an external organization, not one of the ordering of a growing
> experience, one that involves . . . the whole of the live creature, toward a
> fulfilling conclusion.[4]

The Greeks had two words that still inform our language, *poiesis* and
techne. We translate both as "making." But lost is the context allowing
them to be distinguished. Poetic making cannot predict what it will
achieve; as it evolves within the Whole of things, intrinsic value reveals
itself unpredictably. Technical making typically can predict the ends it
will achieve, but cannot assess their intrinsic value—nor their far rang-
ing consequences, many of which are unintended. Once the makings
are distinguished, we can ask whether certain technologies might wed
with true art forms and contribute to constructive and empowering
myths of living.

A Trap

This chapter is titled, "Technology As Ecstasy: How to Deal with It?" In
other words, can we have the benefits of technology without its addictive constrictions and concealments. Ironically, however, in the
phrase—"dealing with technology" thinking easily loses itself in calculating means to ends. Looking for the means to deal with technology
so that it doesn't become addictive can itself be addictive.

Addicts of all sorts confine thinking to clearing the path between
fixes. They confine themselves to calculating means for solving problems. But problems themselves are addictively defined as whatever ob-
structs the next fix. The notions chase their own tails. And they
shrink—"satisfaction" or "fix" and even "ourselves"—so as to conceal
the options they are concealing. The whole framework shrinks, and
ourselves locked inside it.

We need a mode of proceeding that is neither quiescence nor problem solving in its contemporary sense. All current words fail at this

point. Can we speak of "non-action"? Or of "contemplative" thinking? Or "meditative"? But all these words are simply the denials, the reverse side, of the addictive modes I want to avoid. By opposing themselves, they attach themselves.

So I will cautiously explore approaches to an art of living (a *poiesis*) in a rampantly technological age (*techne*), but without fixing this as an end to be achieved by such and such means. In the fullness of time some large or small release from techno-addiction may suggest itself. And it may not. Technological arrogance may be inescapable.[5]

Primal Technology—the Dance with Nature and Its Disruption

During long ages of premodern technology, the handicraft artisan used different tools at different stages in producing an article. Moreover, production tended to be seasonal, at one with the rhythmical matrix of Nature. Spear points, say, were crafted mainly during the hunting off-season. Up to current times in certain peasant communities, the cross-quarter day of Imbalc (half-way between winter solstice and spring equinox) celebrated the production or maintenance of tools that would be used in subsequent months.

Modern industrial technology shows a radical shift away from the seasonal, from natural cycles of exertion, rest and repair, re-exertion. To maximize output in the production line, all tools are used all the time and workers are specialized, typically using only one tool all the time. Numbing repetition suggests a degenerated addiction in which even passing pleasures are missing.

Henry Ford hailed his production line as a way of making a car that could be purchased by workers and farmers, allowing them to escape their isolation. To some extent this has happened. But solutions propounded within corporately addictive productiveness and its paradigm create difficulties that the paradigm does not predict and cannot cope with. Crowded superhighways produce a new isolation and typically degrade the regions to which they give access.

New mitigating options seem to be available: superelectronic tools of communication and large stretches of apparently leisure time, or new possibilities for solar energy and the repair of the environment. Such horizons are sketched by Herman E. Daly and John B. Cobb in *For the Common Good*. But they predict a race against time as conventional economics fails to take into account long-range and widespread environmental costs of "productiveness." Economic development ushers in an era of uneconomic growth that impoverishes rather than enriches.

Acres of arable land diminish as population mushrooms. It is not hard to imagine what will happen when population growth outruns food supply in many areas of the globe. China's draconian measures for controlling population probably presage violently repressive governments coping with rampant misery.

Modern Technology: Tacit Asceticism
The heedlessness and blindness of modern technology can only be explained if we suppose that among its unwitting assumptions is tacit asceticism: the residual male myth that we are not animals, but pure minds that step outside Nature in thought, and heroically control and subdue it in whatever way we wish. Note well Descartes, the well-named "father of modern philosophy" in *Meditations* . . . :

> On the one hand I have a clear and distinct idea of myself as a thinking being and on the other hand I possess a distinct idea of body as an unthinking thing. Therefore it is certain that I am entirely and truly distinct from my body and may exist without it.

This is no stray rumination of an arcane soul, for the words epitomize the secret and perverse spirit of modern materialism and technology reigning nearly unquestioned until very recently. To think that one is a pure, self-transparent consciousness may seem angelic possession, but insofar as alienation from our bodies and from kindred in Nature ensues, we are demonically possessed or infested.

Residing uncritically within modern technology since the seventeenth century is the archetype of patriarchy. Identity of person is construed atomically, in terms of what the atom excludes. But now the exclusion is pushed as far as it can go: each is a mental singularity that excludes even its own body—at least in thought. Can we imagine a woman concocting Descartes' philosophy?[6]

To deny some of the body's primal needs is to invite substitute gratifications and addictions. Of course, achieving a sense of our significance is also a primal need, and its satisfaction may compensate for the frustration of other needs. But can rampant technological prowess and great cash profits compensate for needs left unsatisfied, particularly sacramental needs and propensities (whether acknowledged or not)? There is evidence from the recent history of environmental and human degradation that technological ecstasy that takes over lives is addictive and degenerative in the long run. Some numbers of people are beginning to see this, but whether enough to tip the balance in the corporate organism of the society is questionable.

Given technology's fascination, implicit intrinsic values, and tacit, weird asceticism, we might say that it fast becomes the first world religion. Its assumptions spread over the globe: that endless economic and technological growth—measured in the crassest terms—is unquestionably good. "But what about unrecyclable wastes? The radical degradation of soil, vegetation, water, air? The dispiriting of human beings—and not just indigenous peoples who are utterly lost in modern surroundings?" These questions are ignored by those caught up in the trancelike addiction of power for more power for more power. Even rude observations are crowded from attention, particularly this: that while a few peoples or castes get richer, the vast populations of the rest of the world get poorer.[7]

Some predict that very early in the next century two-thirds of the world's population will live in cities. The greatest migration in human history is in progress, and ever greater homelessness, rootlessness, and despair: people living in, on, and from garbage dumps. This is a hunting and gathering more wretched than our forebears could have imagined, for they had their array of well-honed skills, their self-respect, their ties to an ever-regenerating, not a dead, world.

A New Art of Life?

The ecstatic thrusts of modern technology must somehow be modulated to fit the ecstatic skills and involvements we inherit from an immemorial past if an art of life, personal and communal, is to emerge. Recently signs appear that this is possible, but very difficult. As without technology we would not have our present crisis, so without it we will not be able to find our way out. Dilated body-selves might begin to relocate technology within another myth more attuned to the full gamut of our primal needs and excitements.

Radical reorientation is needed. Take the current technology of energy production that excavates Earth, bringing to the surface ancient vegetable and animal fossil remains in fluid form, burning them to power engines of transportation, production, consumption. The wastes of the burning of oil are thrown up in the sky. It's as if Earth were a gigantic cow, digesting its decaying vegetation, kicked into violent regurgitation.[8] In about two hundred years drilling, mining, and cutting have transformed the planet and all living things. This technological embodiment of ecstatic activity has acquired such momentum that

many simply assume civilization can mean nothing aside from economies of ever more production and ever more wastes.

*Major sources of energy production and manufacture are all located at remote
spots on the periphery of Australia. So manufactured goods and fuels must be
transported to the inland stations, and certain raw materials transported out.
For the sake of efficiency, "road trains" are used: huge trucks with perhaps four
equally large trailers. They come at high speeds down the two-lane "highways"
shrouded in roiling dust, their momentum so great they brake to a stop only very
slowly. Due to their length they can turn only slowly and with great difficulty.
The motorist approaching them is well advised to pull to the far side and slow
down, prepared to move to the shoulder. It is not reassuring to be told that the
drivers of these road trains sometimes drug themselves to combat boredom.*

*In sharpest contrast, another form of energy production is beginning to be
used in the incredible expanses of the Outback. Occasionally seen is a small
concrete structure and next to it a set of solar panels on a pole to power something—the germ of a very different way of organizing civilization.*

To re-myth our lives so that alternate civilizations can be imagined—
their possibility fed into our whole body-selves—that is the greatest difficulty. Ways of prudently slowing and turning the road train without
stopping or crashing it must be found. Needed: a functioning community of therapists, farmers, capitalists, ecologists, engineers, various
technologists, economists, artists, politicians, both theo- and thealogi-
ans, philosophers, and a new breed who might be called imagineers.

We can at least try to imagine new wholes, corporate individuals, new
ways of coordinating and interweaving elements so that parts feed into
each other regeneratively. To take a comparatively simple case, how
can conservation groups be coordinated? Elements that seem clearly to
be incommensurable might be represented within new contexts and
hitherto concealed interdependencies detected. For example, how
could we compare an experience of the sacred and, on the other hand,
a tally of money profit? Isn't it like comparing apples and oranges?
But an enlarged context reveals many connections, for instance, money
profit is a tacitly sacred thing for many people today, or, for another
example, the devastation of Nature, may, in the end, dispirit working
people to the point at which even money profit is diminished, and so
on.[9]

We can at least ask whether rigid imposition of forms can be abandoned, and lives redesigned as we move along within the vastest hori-

zons. Can *techne* and *poiesis* feed into each other? Can what is new emerge nonirruptively?

In 1956, responding to the question, "Are non-violent and non-destructive industrial civilizations possible?" Peter van Dresser was not pessimistic: "At long last a good deal of scientific and engineering attention is being focused on the problem of effective solar-energy utilization." He went on to sketch a vision,

> Controlled violence is the essence of contemporary mechanism; the violence of fire under forced draft, of exploding gases, of superheated steam, of high-voltage electricity, and bombarded atomic nuclei. This is the technology of Vulcan, closely allied to, and often arising directly out of, the demands of war . . . A future solar technology would appear to be largely free of this core of violence. Utilizing fluxes of low-gradient energy controlled by diurnal and seasonal cycles, its processes would of necessity be akin to the processes of plant growth. . . . A greater intimacy with, and respect for, the subtle phenomena of the natural world would be enforced by the geographic dispersal of the homes and work places utilizing solar energy, and their necessary affinity to climatic and celestial rhythms.[10]

The boughs of trees nod familiarly to us in Emerson's vision, and, in van Dresser's technology intertwines harmoniously with Nature. That so little has been done in over forty years to realize the potential of solar power is an indictment of North Atlantic civilization. We now know what van Dresser could not: of the depletion of the protective ozone layer, nearly incredible wastage of irreplaceable topsoil brought on largely through petroleum based fertilizers, probable warming of the Earth, etc. Perhaps the juggernaut of bottom line capitalism will not be redirectable.

But in the spirit of William James it is reasonable to think that it might be. For if we don't assume so, it surely will not be. What if technological genius made solar power a bit more efficient? Uncentralized and fairly cheap, it would decentralize and dehierarchalize life to some degree, allowing people to leave cities if desired. It would slow us down. Serious gardening on an acre or so of land might become an option for some.[11]

Fruits and vegetables could be raised to supply part of the food, and animals run and pastured. Life harmonized with the seasons might return to some extent. The devastation of wilderness to produce hydroelectric and coal power would ease. People whose livlihood is gained through inscripted communications could communicate through computers, modems, e-mail, faxes, the Internet, and the need for daily commuting might cease. Life could become more like a big circle instead of a pyramid, a circle that might resonate within the Great Circle.

But more than moving to the country is needed; it may not even be necessary for many. Euro-Americans can move to rural surrounds and still bring with them a deep alienation from this continent. It is not *our land*, it is merely something to be exploited for immediate, crass gain. In a brilliant study, Carol Lee Sanchez discovers attitudes of respect for the universe common to all tribal societies.[12] Knowing we cannot simply return to any particular tribal group, we yet might teach ourselves to live immediately and reverently in the living and non-living world around us. We might teach ourselves that living is possible and desirable at a pace other than a dead run. In a daring stretch of imagination and empathy, Sanchez writes,

> I believe it is time to create new . . . ceremonies that include metals, petrochemicals, and fossil fuels, electricity, modern solar power systems, and water power systems . . . it is very important to make sacred, to acknowledge the new ways and elements in our lives—from nuclear power (which is buried in our Earth and activates the Sun) to plastics and computers.[13]

Can we just "make sacred" all this? It is a breathtaking and perhaps daunting challenge.

The overriding task is to relocate technology within a new mythic awareness and art of life in which we learn how Nature responds to what we do, and we correct what we do next: a self-correcting circuit within a vast circle of beings. We need a visceral involvement in cosmic democracy.

New Friendly Technologies

Technology's marvels—re-experienced within new corporate bodies and more commodious ideas of truth—may yet save us. The reach of all technological advance is toward ecstasy; its grasp is questionable. As we near the close of the millennium, important technological innovations appear at ever increasing rates. Not all these are wedded mindlessly to production, consumption, and the wasting of Nature and ourselves. Some emulate regenerative celestial and climatic rhythms.

John and Nancy Jack Todd have created "organic machines," prototype buildings, for example, that simulate organisms.[14] What does a fabricated system excrete? How do we feed this back? One of the answers is the Solar Aquatic Wastewater Purification System. The sun's energy purifies recycled water. The Todds attack head-on the implicit patriarchal dogma of industrialism: that we lock an identity into an envelope and control it by pouring energy into it and casting its wastes aside. What if identity is a field—an area of reciprocating interaction across a membrane? What if a fetus floating in the womb is a better metaphor for individual identity than is a billiard ball on a table?

Michael Terman et. al. have developed a technology that precisely simulates daily waxing and waning of natural light for use in bedrooms where natural light is restricted. We hear of "twilight therapeutics" and "light treatment" used to cure sleep disturbances, dysrhythmia, dysthymia, or outright depression. Patents have been issued (1992) to the Medic-Light company for the Naturalistic Illuminator System. These are not cure-alls, but they are very significant.[15]

Paul Ryan, polymath, suggests ways in which TV might be used to lure viewers back into the rhythms of Nature, at least to some extent, through the use of "fixed remote cameras that observe tides, sunrises, sunsets, heat patterns of the city (infrared cameras) etc." Ingeniously, in his "Art and Survival: The Earthscore Notational System for Sharing Perception of the Natural World," he draws an analogy between the patterns of Nature and a musical score:

> A notational system designed to interpret the natural world must . . . be based on clear "notes" elicited from the natural world. For example, in order for videographers to record salmon spawning . . . they must understand the "figures of regulation" guiding the "performance" of the salmon . . . then the videographers are in a position to scan the ecological system for perturbations and alert us that something might be disturbing the underlying figures of regulation . . . it is possible to shape television so that it will constantly be singing nature's score.[16]

Ryan has even developed schemes for financing the work through WNYC, the New York public radio station (the financial woes of the city have nipped this "nonessential" in the bud).

The weary addictive ecstasies, the road trains, of long-established systems of production, consumption, and waste can be countered only by a new ecstasy allied with Nature, facilitated by technology, and inspired by a kind of cosmic sense of fairness and democracy. With encouragement this new possibility for ecstatic experience might excite significant sectors of the population.

Many-armed technology reaches out ecstatically in many directions, from emulation of Nature to attempts to eclipse or replace her. We don't know, really, what will happen, for technology extends the labile **spiritual body that moves unpredictably. As I write, Jaron Lanier commands spots on the mass media touting his invention of Virtual Reality.[17] This technological marvel of communication challenges impishly our ability to assess its value and plot its course.

Virtual Reality—the Siren Call

For a considerable time the computer has been mated to the nervous system, in word processing, for example. But for Lanier this is Stone Age technology. Screens are now built into eye phones that fit snugly around the eyeballs so that, as he says, "You're a part of the scene." We think of Jerzy Kosinski's novel *Being There*.

Within this technology the body-self's habitual ways of orienting itself within itself and contrasting itself with the scene-viewed are considerably suppressed; individual identity is stretched, perhaps strained. In addition to eye phones, wired gloves are worn that simulate tactile sensations. Even a full bodysuit can be donned—all controlled by computer. Lanier can, he says, "map your arm into a gazelle's leg." Or the computer can be programmed to send images of participants—parted by thousands of miles perhaps—into each other's screens (and "tactility"). Adventures of "interaction" can ensue. Perhaps both persons will be programmed with gazelle's bodies? "The physical world resists us utterly," observes Lanier, so why not circumvent it and allow our imaginations full reach?

Such innovations and ones ever more ambitious seem inevitable. Humans need ecstasy. As James knew, we crave moral holidays. Lanier speaks of trading eyes with one's electronically linked partner, of the funny sensations of sharing a body.

There are possible virtues of Virtual Reality. Perhaps those partially paralyzed can build up substitute skills through the elaborate feed-back system. Or perhaps those sexually impotent for "psychological" (**spiritual) reasons can experience such sexual "success" in Virtual Reality that inhibitions dissolve when off the machine.

The dangers are obvious: that VR will become a substitute for ecstatic engagement in reality, a massive addiction that diverts attention from big personal, social, ecological questions. Questions such as, Can our needs for bread and for ecstasy both be met within a sustainable future? VR used mindlessly contributes mightily to what LaChapelle calls the

International Growth Society's corporate escapism and addictedness. It is an ultimate consumer item: needs for entertainment—substitutes for risky exploration—are generated and satisfied, needs that quickly reemerge and must be resatisfied again and again.

Lanier concedes that the "images lack preciousness and easily become stale." I am glad to hear this but not surprised. For, short of delusion or insanity, participants know they are experiencing artificially produced images only, not presences or images of *real* things. But, of course, VR images gone stale may prompt withdrawal symptoms so strong that only another fix on a new version of VR stimulation is desired.

Real things: things that overflow, fill and light up within us, swell, move on their own, perhaps in ancient rhythms, and—if compatible—refresh us. Real things generate presences and images that cannot be completely controlled or even exhaustively inventoried. Real things are endlessly overflowing as long as they exist at all. We can address and be addressed by them, knowing they have a life of their own. We can count on them when we walk on the Earth, or stand stock-still with awe, or slide into sleep. Primal needs are fundamentally real ones, and they can be satisfied only by real things. Persons who consult counselors complaining of lives that are empty and flat, pointless, hunger for the experience of reality in its ecstatic fullness. VR will not satisfy for long.

But the lure of immediate gratification is uncannily strong, as is the inertia that results from achieving it. Mihaly Csikszentmihalyi does well to wonder why people spend so much time watching TV, for example, when his research shows that they actually enjoy doing other things more—working in a challenging job that fits their capacities, for instance.[18] The answer is the insidious pull of addictions, the vitiating effect of dependence upon immediate gratification. One slowly sinks into what Dostoevsky in *The Gambler* calls the mire. Like infants, we demand the rewards without doing the work.

If we are coherent and vital to the extent we feel ourselves vital parts of the real world, where does technically assisted narcissism and dissociation leave us? I imagine orgasm in these circumstances would be experienced as merely intensely pleasurable but ephemeral sensations . . . unless we were to be utterly deluded. Given the restless progress of technology, some will find out for sure soon enough.[19]

Face to face accountability is missing in cybersex. This absence might be welcomed at first as relief. But if sex of this kind takes over one's life, the relief will have to be repeated addictively. For the ecstasy of

responsibility and the momentum of real life will be missing, and withdrawal symptoms of flatness and vacuity will emerge and demand to be
counteracted *now*, and again and again. W. E. Hocking:

> To be able to give oneself whole-heartedly to the present one must be
> persistently aware that it is not all. One must rather be able to treat the
> present moment as if it were engaged in the business allotted to it by that
> total life which stretches indefinitely beyond.[20]

Ultimately there is no escape from the one reality that enwombs us.
But when we isolate and eccentrically emphasize certain interests, dis-
tortions occur. We may lose the sure path of our continuous central life,
indeed, our life's moral core. That is, the rooting assurance that it is
best for us to safeguard an awareness of reality that is as honest, firm,
true, comprehensive as we can muster. This must be an intuitive assurance and responsibility antedating all theorization and manipulation,
must be a stabilizing ground mood.[21] It must grasp us as unconditionally as myth once did.

Coordinating Mythic Power and Technical Expertise
Electronic media, particularly ubiquitous commercial TV, are immensely difficult to put in perspective, because they insinuate themselves within our perspectives and control them. When inanity,
banality, violence, irresponsibility are broadcast, these inflate and become paradigmatic simply by being shown. By increments too small to
be noticed at the time, TV's flashing images have carried us into uncharted waters and numbing trance.

Aristotle observed that rational persons wish to be recognized and
respected by those they respect. But most of us are far from reflective
and rational much of the time. Any medium that picks persons out of
the mass, and multiplies individual or corporate images untold millions
of times, typically overwhelms critical judgment and extorts respect.
Afoot is a new mythic force, a vast witnessing presence, a corporate
body with transfixing countenance that mesmerizes the "neutral" body-
self. It is addictive infestation of the corporate body.

Media stars have their images multiplied myriad times by TV and
broadcast in every direction through (and out of) the atmosphere:
Mickey Mantle, Curt Cobain, O.J. Simpson, Princess Diana. They are
inflated to quasi-mythic size—gaseous giants around which new national and international "communities" form amid the rubble of traditional cultures. The distinction between viewers and viewed smears

somewhat: viewers identify with giants viewed. We feel vaguely author-
ized and recognized by them (vaguely and abortively, for they are not
aware of us at all, and we dimly know this). Icons-and-icon argot, a
common language, arises and perpetuates itself for awhile. When particular giants grow stale and withdrawal sets in, new luminaries must
appear, new centers for charged and ephemeral communities.

Traditional rituals of passage and recognition rhythmically interwoven in Earth and space-time restore experience of the sacred and
build coherent identity, for they meet a primal need: to be recognized
as a real participant in a real group that believes itself to be a recognized and valued corporate member of the Whole. In lieu of these, we
formulate rituals ad hoc, and are carried about in their shifting winds
and dust. Some turn out to be tedious personal and corporate addictions, others the sacrificial violence of mobs. Sacrifices to questionable
gods and heroes to cure unspeakable flaccidness and loneliness.

*I was walking down the main street of the university and noticed a man walking
in the same direction about ten paces in front of me. He looked strangely familiar, but I could not say who he was. Dressed in a fine pin-striped blue suit and
carrying a chic briefcase, he was slightly hunched and inflated and swung his
arms like an ape. I hurried along, and when I was about fifteen paces ahead of
him quickly glanced back. It was that man all right, that young professor of
history who had just been made a full vice president of the university. His whole
demeanor, his swaggering walk, his whole way of being a body-self, had so
massively changed that I marveled.*

*And just the other day a professor stood in front of me at a reception. He was
no more than ten feet away. Apparently he was someone I knew, but so altered
that I reached for my glasses to better assess what had happened to him. It
assuredly was he. But his face so grave and weighty, his body fixed to the floor,
I recognized him with a shock. I had seen him a few months before. Now his
boyish smile had vanished, replaced by a supercilious gravity. What had caused
this? I engaged him in conversation. His book had gotten rave reviews, had
already sold twenty thousand copies, and he was "traveling around the country
giving talks about it." I was greatly impressed and envious. I did not bother to
find out what the book was, nor have I had it in my hand to this day.*

*Both these academics looked drugged. Unlikely it was, however, that they had
ingested any substance that could be labeled narcotic. We were all drugged with
fame.*

Media Stars: New Demigods and Demigoddesses

Contemporary communications technologies disconnect circuits of meaning making and life formation evolved over millions of years within local environments. We learned to form ourselves as persons on the basis of responses from physically present and known beings who knew us as persons; also from mysterious gods and goddesses. Communities of persons rooted together and stabilized over time through ritual are now scarce. The very concept of place or locale is drained of much of its meaning. How, then, can we echolocate ourselves?

Known (however superficially) by so many unknown (or superficially known) people, how can I be known to myself? Stars of electronic media write large the average person's dilemma. We tend to identify with the stars as does a voyeur with persons seen through a keyhole. Differently put, they are the heads of corporate bodies that include us shadowy peripheral members. They carry us about. They live and suffer for us. Fascination with their fate is abortive and consuming self-concern.

After drinking much of his life away playing the part of the baseball hero MICKEY MANTLE, Mickey Mantle repairs to the Betty Ford clinic. Emerging sober he says, "For all those years I lived the life of somebody I didn't know. A cartoon character. From now on, Mickey Mantle is going to be a real person."[22]

When the young rock star Kurt Cobain killed himself "at the top of his game," the *Village Voice* reported, " 'Kurt Cobain is not a person,' says Daniel House, owner of Seattle independent record label C/Z. 'He's turned into something that represents different things to different people.' " The writer, Ann Powers, goes on, "In our century, 'fame kills' is almost a mantra; add Cobain's name to the pantheon. . . . But it's hard . . . to pinpoint the moment when a star like Cobain slips into that nether realm, becomes flat and reproducible, something read instead of something known."[23]

Either known by those we don't know, or knowing those who don't know us, we slip into the nether realm. Watching TV, we know fine lines on faces of people who don't know we exist. The self diffuses in loneliness.[24] It does not share itself and return to itself augmented and coordinated through others' recognitions. Circular power returning into itself is short-circuited, can no longer be self-corrective and revitalizing. Our thought, incarnated technologically, no longer coordinates with the ancient self-moving organisms that are ourselves. We succumb to baneful trance.

A Rough Sketch of Our Place in Nature

How do we wake up? What is our place in Nature? That is, how do we realize our own truth as beings formed in Nature, who modify it, to be sure, but always within it?

In a climb up a steep rock slope, in splitting a pine log, cultivating soil for planting, turning metal on a lathe, in all such activities we encounter the sheer *thatness* of things, their reassuring resistance. Without it we would be disoriented; there would be no support at all during disasters. All things, along with us, play a part within an awesome Whole that has a source and ultimate cohesion we cannot explain but may feel. If there were no Whole, things that are clearly precious could not exist. Each thing is considerable just because it is a being in its own right, even a bubble is. We can glory in it, address it in a sparkling way as a fellow being.

Here is a bird's eye view of our place in Nature, of the many facets of our need to *be*:

(I) The inorganic world is valuable in itself. And valuing it for itself is an ecstatic, self-evidently valuable experience for us. Scientists now conjecture that the daily cycles of light and dark have structured the regenerative cycles of every cell in all living things.[25] Without the inorganic translated into the organic nothing could live. All life evolved out of the inorganic, and at some point will be swallowed back into it. And we are dependent upon the inorganic not only for bare physical survival. Habitual warmth or cold, light or dark, wet or dry evoke typical ways of vitally communicating and being in human cultures. Ritual observance of cycling seasons composed the initial fabric of all art, religion, civilization.

(II) We share with other animals sensuous solidarity. These are self-evidently valuable activities for all of us: to move freely, drink good water when thirsty, rest when tired, eat nourishing food when hungry, experience a variety of sensations, communicate primal states of awareness, cohabit with others of our species, and appreciate the infusing power of other species—to touch, caress, and perhaps kiss kindred beings. Animal bodies interacting with the surround are not just objects, but subjects giving meaning to us and the rest of the world. Nonhuman animals live within domains of meaning co-created by them and the world, even if they cannot talk or think about this. Many do think—at the very least about what is required to satisfy immediately felt needs. (How much of our instinctive and impulsive meaning making are *we* aware of?) Consciousness need not be reflexive, that is, be about itself—although some nonhuman animals seem to possess a degree of this. Though we will probably never speak adequately of nonhuman beings' experiencing, the shared world of earth, sky, air—the world's presence—allows us to be convivial with them. Extended family knits together all living things.

(III) We need and can have sensuous solidarity with all animals, but we need and can have *con*sensual solidarity—in explicit and deliberate forms—with other humans. Using shared symbol systems, we have aimed traditionally to grasp the Source of all things and the world *as* world. Origin stories have framed us in relation to Source and placed us in the Whole. Such myths of unconditioned and unconditional reality and value have contained all other ideas within themselves, and exercised an enormous influence on the way we have lived on this planet. True, I alone move relative to something and it appears as it did before, or I look at something that appears to be right there and touch it and it is there—its invariability is evidence for its reality. Or, equiprimordially, we move about in each others' physical presence and jointly confirm things or trust each others' confirmations. All these are activities not much different from those of nonhuman animals. Beyond all this, however, is distinctive human language and far flung systems of shared symbols. Today, awareness expands explosively beyond any individual's, or group's, direct confirmations. We assume that some people somewhere sometime have varied their positions or postures or transformations, and what they believe to be there remains invariable, hence is a reality that would exist even if we ceased to do so. Our feeling of directly participating in, contributing to, and being obligated to the human community and to the Earth becomes ever more tenuous. Today new modes of symbolization embodied in technology mushroom, while older modes tied to myth and ritual shrink. We possess electronic and atomic tools, appliances guided conceptually and symbolically which work at great distances to alter Earth and ourselves beyond our early ancestors' ability to have imagined, and beyond our own to reckon the consequences of what we're doing. Never before have we possessed so many possibilities and so little sense of priorities, individual or communal. Our ability to produce ecstasy-laden endorphins in distinctive ways easily becomes wayward or frantic activity. We seize possibilities and can hold ourselves responsible for achieving what we think should be, justice, say, as idea and institution. Or, vast wealth and power for a few at the expense of the many. We form an ideal of intersubjective agreement and truth, but, abstract as it is, we easily depart from it. We can grapple with our freedom and responsibility, or distract ourselves with mind games and evasions. Our primal need to *be*, and to feel excitement in it, has become exceedingly complex, and our difficulties are perhaps intractable.

Our Crisis

Thus our crisis today is that we find little or no foundation for (III) in (II) and (I). Distinctively human powers of symbolization, self-consciousness, life shared through technology, and so on, have lost their bearings in the sensuous matrix, and in the local and cosmic surround, and so we rattle around in nihilism, unmet primal needs, assorted addictions. We aren't confident of what's good and right, are not members in good standing of a vast community of beings, addressing and addressed. The consensual domain shrinks—into legalisms, for example. We can't seem to see that value resides in the basic human needs and their satisfactions, with a need for a feeling of our own significance being paramount: can't seem to see that a fundamental core of what's really valuable is what remains invariant from various responsive and open points of view. Lost is the whole Matrix essential to human life, body-self's **spirituality. No longer does art consistently play its pivotal role connecting the sensual and the consensual. The sense of the sacred deserts us: the sustaining numinous experience of body-self and local environment nesting in the everything else. Technology extends us in ways we think we understand, but limits us in ways that conceal themselves. Unlike our hunter-gatherer forebears, *we no longer consistently integrate our freedom of movement as animals and our freedom of thought as humans.* Despite stunning technological achievements, our primal need to *be* fully and coherently is not met.

> *Turning and turning in the widening gyre*
> *The falcon cannot hear the falconer;*
> *Things fall apart; the centre cannot hold;*
> *Mere anarchy is loosed upon the world.*[26]

Technological Extensions of Our Bodies?

But there seem to be examples that counter the claim that we no longer integrate our freedom of movement as animals with our freedom of thought as humans. When the space module *Pathfinder* landed on Mars in July 1997 and released its rover *Sojourner* to sample that planet, some called this a 119 million mile extension of our nervous systems. Over a million viewers tuned to the NASA web site saw Mars through the eye of a camera mounted on top of the rover as it crept over the rocky surface formed billions of years ago. A miracle of technological mastery—especially twentieth century computer-power—cyberspace used to open up real space and time for our experience. Isn't this the pinnacle of *integration* of movement of thought and our bodies? Millions stood—or sat—in awe and exulted. A basic need was being satisfied dramatically: to explore.

Still, we might think further. Though none of our senses function in total isolation of the others, vision is greatly dominant in the case at hand. As in computer use generally, trunk activity is minimal, hearing limited to the human voice commenting on the scene viewed. Watching TV we lack even the tactility of finger tips brushing keys. Moreover, the experience is episodic, easily growing stale through overexposure in a few days or weeks time. David Bohm notes that we can sustain experience only within a totality of significance, that is, only within a felt sense of the world as boundless, and of our body-selves as tiny but vital parts of the family of beings, emerging, enduring, and passing away within it.[27] As prehumans and humans took shape through habitual adaptations over millions of years, so we both inherit their proclivities and take shape through individual and group habituations over mere decades. Habits acquired in modern life and those from archaic times jar. The long runs tell, not the quick fixes, and here the rule holds: in the computer age movements of thinking tend to be uncoordinated with movements of whole bodies.

If appreciated within the horizon of a universe that exceeds our powers to imagine, computer technology can contribute greatly to our sense of the sacred alltogetherness of things. But fascination with it easily constricts attention. We lose a sense of the Whole, and our body-selves as wholes within the Whole. We become feverish eccentrics.

Healing Arts
The only healers for tens of thousands of years were shamans. It is nearly impossible for us to understand how they succeeded, when they did so. If we try to understand the twentieth century's fragmentary "psychosomatic" medicine, we typically think of ideas and beliefs some-

how acting beneficially upon the body of the thinker: thoughts as invisible billiard balls.

But this notion is just another manifestation of our perpetual unease, dis-ease, crisis: we do not integrate the movements of our thought and the movements of our animal bodies. Ideas or beliefs don't have a separate mental reality that could act miraculously-mechanically on the body. Regardless of what we fancy the words "ideas" and "beliefs" mean, they really refer to just an aspect of the single system of energy that *is* ourselves, an aspect that runs throughout the system, which exhibits another aspect that we call our bodies. Body-self believing something about itself and its place in the world alters itself. When we habitually divorce movements of thought which involve only portions of the nervous system from movements of the whole body, we are at odds with ourselves.

Claude Lévi-Strauss has detected certain parallels between the practices of traditional shamanic healers and contemporary psychoanalysts.[28] I repudiate his ever-clinging Cartesian mind/body dualism, but appropriate a living kernel of his insight. Psychoanalysts try to achieve a deep working relationship, a "transference" in which clients transfer to the analyst early feelings toward authority figures, and a "countertransference" in which the analyst cares for the client. But the analyst—in contrast presumably to the client's early care-givers—responds in a way that respects the primal needs of the client and his or her situation. Typically, psychoanalysis has been lengthy, for it takes a long time to work through a transference-countertransference, so that clients become mature enough to test reality for themselves in the present.

Psychoanalysis has emerged to fill the vacuum left by the loss of shamans, and the vitiation of social and religious institutions generally in North Atlantic culture. Lévi-Strauss cites a pregnant woman with a blocked delivery who has appealed to a shaman, in this case a man. With the greatest patience he enacts traditional "spirits" who enter and exit her room. As the woman fixes trance-like on these passages of spirits, she begins to be diverted from the pain that possesses her and blocks her delivery. Her womb and birth canal, apparently, begin to realign themselves in normal circuits of entry, conception, gestation, exit—delivery. In this case at least, the treatment works. She manages to give birth. If we think of what we are as multi-aspectual systems of energy within enwombing circular power and consensual life, the cure is no longer so surprising.

Now, do we betake ourselves to either psychoanalysts or shamans? We might, and one or the other might work, and neither might. For psychoanalysts, formed in the same deracinating North Atlantic culture that formed the rest of us, may be unable to enact "the good enough parent" with enough skill and patience to actually get their bodily presence into the client's body. Analysts may remain stranded in talk, contemporary symbols and signs, theirs or the client's.

And shamans, equipped with their drums and perhaps native hallucinogenic drugs, may be unable to get through to clients deeply inured to modern technology, who are unable to fix belief on the powers of the shaman's body, its exfoliating envelopes of energy—photons, pheromones, sounds, words, gestures, etc. Or who cannot respond to whatever "spirits" the shaman invokes.

Individuals' Initiatives and New Arts of Life

But perhaps we remainders are not incurable. Disequilibrated by great stresses, the body tends automatically to regain balance. If we allow this, it can happen way beyond the scope of Walter Cannon's findings in the 1920s. We might again find a world.

Things can be perceived and characterized only in terms of how they pour through our expectant bodies and gain a response. The most elaborate tools and perceptual aids must deliver stimuli that can interest and be interpreted by us. Events may be occurring in and around us so unimaginable and unexpectable they cannot be perceived at all. The only response is numbness. But a body-self embracing the possibility of healing is greatly different from one that does not. Open to nameless regenerative possibilities of the world, the body's own regenerative capacities must respond. In some cases openness makes the difference between life and death.

Neuroscientists now know that the body is not composed of separate systems—the immune system, say, or the higher cortical systems, or the emotion centers. Hormonal "messengers," neuropeptides, flow in the blood throughout the body, so that what we believe and feel is not separate from the ability to fight disease. Interfusing with environments, the whole body-self knows and feels and can heal or infest itself, sometimes dramatically. A scientific field emerges hesitantly: psychoneuroimmunology.[29]

The most insidious of technology's dangers is this: Drastically enhancing some of our powers—particularly for direct dealings with immediate effect in the public world—other powers languish. These are

**spiritual ones: freedom to turn one's gaze in a new way—perhaps in the aloneness of the night—the ability to hope. We can hope that scientists can hope, that their science will continue to expand and limber-up, and that technology and hardened attitudes can do the same.

Hunter-gatherer ancestors not only knew what to expect when and where, but were open to the unexpectedly encountered. They were intensely alert to anything happening anywhere around them. They were eager to live.

Art-like or religious-like rituals enacted in the most unlikely place—cities bursting with computers—might begin to reconnect us to archaic sources. Ways of being that are irrelevant to city life as we commonly manage and live it are essential to us products of millions of years of evolution. The wilderness way of seeing is what trackers of animals and birds call splatter vision—alertness to everything on every hand. The merest fleck of something seen out of the corner of the eye is not ignored but noted. The body-self quivers with alertness: *movements of thought and body are one*. Thoreau:

> But it sometimes happens that . . . the thought of some work will run in my head and I am not where my body is—I am out of my senses. In my walks I would fain return to my senses.[30]

The wilderness way of being, retrieved ritualistically even if apparently only personally, transforms cities into something more than the humanly calculated and contrived. Commonly conceived, reflections of things, say, are adventitious, seldom counting for much. But this is so only when we stay hitched to demands of subsistence since agriculture, and lose footing as organisms formed in wilderness Nature as artist-artisan-hunter-gatherers. We are beings who can roam free: ancient creatures alert to things *and* their reverberations, shadows, echoes, reflections on water and rock—to voices, touches, glimpses, glances of will-of-the-place. We might address things as marvelous, not limited to their envelopes but propagating their being suffusingly. Angus Graham:

> The effect of opening the senses . . . is most spectacular in the city by night . . . in a tumult of mirrors, wide windows, polished metal and plastic, contrasting colors, and glaring lights. Nothing that architects plan has the complex beauty of the astonishing collages, the endless changing mobiles, springing from chance conjunctions to transform the most commonplace building.[31]

To be open to chance is to be awe-struck in a new—and old—kind of way. Thus engaged is a new—and old—ritualistic way of being.

Our Problematic Adaptibility

By adapting thoughtlessly to our own creations we go out of sync with the creature we each have become through multiple ages of the species' adaptation to Nature—dislocated from inorganic and organic nature, our own truth. As a culture, we have not yet learned to make human sense of the reality of evolution. All the arts and sciences must come together with religious and healing impulses from around the planet to form once more an art of life. Again, Walt Whitman:

> Urge and urge and urge,
> Always the procreant urge of the world.

Can we rediscover a feeling for oceans lapping our shores, sea creatures swimming or crawling, birds flying? Wilderness is other than civilization, but is essential to it, and civilization must reclaim its intimate other as its own. This "must" is more than the instrumental or hypothetical, "If we would solve the problem of survival we must. . . ." It is the unconditional and ecstatic obligation to fully be, to be in the truth, to find kinship on the mythic level that is simultaneously actual. Only thereby will our urges be experienced as through and through our own, the threat of addiction dissipated.

We need a multifaceted openness like Nietzsche's creative play, "the innocence of becoming," in which nothing happens for the sake of anything in particular—which is the same as saying that everything happens for the sake of everything. Are we inventive enough to get Nature and its wilderness resonating in our technology, achieving what Emerson called an original relation to the universe?

A Navajo version of the Goddess is Changing Woman. Unlike the transcendent male deity who creates once and for all, Changing Woman creates continuously, and her mythic root might press up and flower through us if we let it.

By multifaceted openness I mean an attitude that is radically open, humble, teachable, hopeful—that is willing to learn from any source, animal, vegetable, human, inorganic (for the mountains themselves are presences), scientific or technological, living or dead, past-and-present or just present—that is willing to start over again. By this openness I mean being swept up dancing with Changing Woman, from new angles seeing new possibilities. Wholeness attracts.

Yeats, in "Man and Echo:"

> *In a cleft that's christened Alt*
> *Under a broken stone I halt*
> *At the bottom of a pit*
> *That broad noon has never lit,*
> *And shout a secret to the stone.*

Notes

1. Quoted in Bigwood, *The Earth Muse*, 224.
2. See Marija Gimbutas, *Civilization of the Goddess*, x, 19, 223: "In all prehistoric art, the vulva is never to be seen as a passive object, but as a symbol for the source of life itself . . . the cosmic womb. . . . Figurines representing the generative forces of the Goddess are always depicted with large vulvas or pubic triangles."
3. See William Barrett, *The Illusion of Technique*.
4. *Art as Experience*, 81.
5. See David Ehrenfeld, *The Arrogance of Humanism* and *Beginning Again*.
6. In fact, I once did hear a woman present a theory of the self's identity that was very Cartesian—an intricate paper that made no reference to the body. Only then did the grotesqueness of the whole project fully dawn on me. (Her body would not be ignored completely: she inhaled her cigarette smoke deeply into it.)
7. For the destructive effects of "development" and "aid" on the local economies of "third world nations," see Daly and Cobb, *For the Common Good*, 18–21, 289–90.
8. An ancient image for the Great Goddess is a cow; for example, Io (Ionia), or Europa (Europe) were totemic or archetypal creatures for the people of these areas. Treating cows as merely matter to be manipulated for profit is symptomatic of patriarchy and **spiritual decay.
9. See Paul Hawken, *The Ecology of Commerce*.
10. "The Coming Solar Age," 363 ff.
11. Wendell Berry, himself a farmer, points out that "most of the vegetables necessary for a family of four can be grown on a plot of forty by sixty feet." *A Continuous Harmony*, 82.
12. "New World Tribal Communities."
13. 352–3.
14. For example, *Bioshelters, Ocean Arks, City Farming*.
15. Dr. Terman has kindly supplied me with helpful materials. I carp at only a few points. Would "winter depression" occur in cultures that did not obsessively divide the "good light" from the "bad darkness"? Would light therapy be even more effective if patients were encouraged to contemplate what was being simulated: Nature's grand periodicity?
16. Paul Ryan, *Video Mind, Video Earth*, 379 ff.

17. From his talk, "Virtual Reality" given to The Reality Club, New York City, February 28, 1990. (Yes, such a club actually existed.)
18. *The Evolving Self*, 42 ff, "Addiction to Pleasure," and 198.
19. See "Cybersex: Virtual Reality Gets Hot," by Johnny Dodd in *New York Perspectives*, January 28–February 3, 1993. "There are no human obligations. . . . Imagine your own sexual robot . . . Technology is going to bring technophiles that much closer to what they want—a pure selfish existence. . . . Most future sex is designed to fulfill your fantasy, without the mess and bother of dealing with another human being."
20. *The Meaning of Immortality in Human Experience*, New York, 1957, 155.
21. See Albert Borgmann, "Artificial Realities."
22. *Sports Illustrated*, April 4, 1994, 77. He had only a short time to live. I wonder if he realized his goal.
23. *Village Voice*, April 19, 1994.
24. Media stars might be regarded as hostages around whom a community can form. Spectators seize on a person about whom they can express themselves to each other. A kind of carnival atmosphere prevailed throughout much of 1994–5, as O. J. Simpson was used to magnetize and entertain.
25. *Science News*, "Biological Clocks Fly Into View," 152 (December 6, 1997), 365; and recall note 8, chapter 1, above.
26. W. B. Yeats, "The Second Coming."
27. "Soma-Significance," 6.
28. "The Effectiveness of Symbols" in *Structural Anthropology*.
29. See Candace Pert, "The Chemical Communicaters." Also in the Moyers anthology: Margaret Kemeny, a conceptually sophisticated scientist.
30. "Walking," 1994 edition, 9–10.
31. In Bockover, *Rules, Rituals, and Responsibility*, 162–3.

Conclusion

The Awesome World

❧

How do you know but ev'ry bird that cuts the airy way,
Is an immense world of delight, clos'd by your senses five?
—WILLIAM BLAKE, ''THE MARRIAGE OF HEAVEN
AND HELL—A MEMORABLE FANCY''

❧

In delighted regard Blake opens to the bird and flows with it and is buoyed by it. He opens ecstatically to its immense world of delight. The most primal of all needs, to *be* fully, and to delight in this, is met.

Gratifications that are substitutes for primal ones are never sufficient and must be repeated slavishly. They are counterfeits and can never be whole-heartedly one's own, one's own truth as a body-self. They are bafflingly "both me and not me." They disorient us, undermine our sense of our significance as agents and persons. We fail ourselves at the heart of ourselves and must feel guilty.

Attempts to kill guilt by denying it and projecting it onto others and other things only exacerbates it. Now it is felt as a pervasive unease and discontent on the margins of consciousness where it is difficult to acknowledge. More dishonesty and confusion is created. The self punishes itself all the more.

At One with Body-Self, At One with the World

Ecstasy withers if Blake regards the "classic five" senses—sight, hearing, touch, taste, smell—as merely inner sensations generated by five sorts of exteroceptor organs located on the body's surface. Trying to get to inner experience by initially objectifying the body is self-defeating. To objectify the body is to detach in one stroke from the bird and from one's own surging, continuous, **spiritual life.

Not firmly planted in itself and in the Earth, dominated today by scientism, body-self takes in thought a strange position outside itself, an automatic abstraction and perverse ecstasy. Fixing on its exterior, the outer sense organs, body-self detaches itself from its other sense modalities, particularly the empathic viscera. But this means it can no longer live in the bird's life. It tends to fill the emptiness with repeated ephemeral sensations, addictive fillers and stop-gaps.

Kierkegaard spoke derisively of "the lunatic postulate": to abstract and then forget what one's done; but not completely automatically, for there is a touch of fearful and guilty mendacity. To reduce oneself to an object in geometrical space that merely observes the moving surfaces of other objects is pathetic and culpable. One fails oneself at the heart of oneself, and other things at the heart of themselves. No amount of addictive "adaptation" and apparent normalcy can completely conceal the guilt.

Primal Peoples

Primal peoples appear to be almost another species. In trying to avoid what we regard as their hardships and backwardness, we miss whatever world of delight they possessed, and try to compensate for an ecstasy loss we cannot articulate. (Even some of their great fears—which we would consider superstitious—did not leave them emotionally empty.) When primal peoples ingest hallucinogenic materials, they do so within symbolizing practices and ritual settings that have evolved through un-told centuries of adaptation within regenerative Nature. The very word "hallucinogenic" expresses prejudice: people are assumed to *hallucinate*, that is, not to perceive reality. Indigenous people enter what some of us, trying to be empathic, call "alternate reality." But the idea we are trying fumblingly to express means different things in the different cultures.

Shamans and their native clients are rooted perceptually and em-pathically in their everyday surround to a degree that the typical secu-lar urbanite does not imagine. If drugs are used, they are usually ancillary; to brand this addiction is simplistic. When allies are sum-moned from "the alternate realm of spirit," they assist by inserting themselves in the everyday perceptual surround. For example, a hunting party loses its way in dense forest. Perhaps assisted by drugs, and tapping memory and imagination, the shaman goes into a trance and receives information about the territory. Or perhaps a person is sick with a disease native to that region, and the shaman invokes curative agents that interdigitate precisely with the group at that place and time.[1]

But the denizen of North Atlantic civilization, equipped with abstract symbols and the wonders of modern technology, has learned to float practically detached from any particular perceptual surround, any local environment with its ages-old regenerative cycles. Either as onlookers or participants in the shamanic ritual, we tend to regard "spirits" (if we regard them at all) in impoverished ways: as animated objects coming from nowhere. Lacking is any identification with them. And after they have riveted attention by working inscrutably, they depart to nowhere. As they are dislocated, so are we.

The shamanic impulse is not simply lost in urban civilizations today. Bodies ever entangle themselves in archaic energies, though unwittingly. For example, advertisers of powerful cars capitalize on potential buyers' identification with animals, their residual, probably unacknowledgeable, shamanlike urge to be incorporated within them without being consumed.[2] In one ad for a car the driver turns the key and the motor "barks through its dual exhausts." Another ad promises that "utter ferocity arrives as the tach's moving display skitters past 4,000 rpm" and "what feels like the hand of God presses you back into the seat." Yet another: "It's not a car. It's an aphrodisiac. . . . It's what happens when you cross sheet metal and desire."

Advertisers embellish heavily, but they know what appeals to many, and experiences resembling those they describe do occur. But typically these are not integrated with a total way of life in a total environment.[3] In fact, cars and all they entail—highways, polluted skies, nearly immobilized bodies—tend to suppress our lives as total organisms coping with periodic demands of a total environment and earning periodic rewards. Ecstatic experiences in cars, indulged at random, tend to split off and fly away.

David Bohm: "Anybody who is self-centered must be divided, because in order to become self-centered he must establish a division between himself and the whole."[4]

Alternate Reality

"Alternate reality" or some roughly equivalent phrase—so rich and relevant for indigenous people—carries a different and confused sense for most urban secular people. Easy it is for trance or "hallucinogenic" experience to lead us even farther from effective contact with local environments and Earth and wilderness and into addiction, which thrives on placelessness, confusion, emptiness, the loss of the sacred.

This is why the topic of "hallucinogenic" drugs and alternate reality has been kept for the very end. I have tried to detect and unpack the extraordinary potentials for ecstasy in "ordinary" experience, to avoid jumping into a "discontinuous" "occult" region before discovering these potentials.[5] As T. S. Eliot wrote decades ago, life for many today is a desiccated world; they yearn for "the sound of water bubbling behind the rock." But the spiritual life gained through leaping prematurely into an "alternate reality" is typically thin, evanescent, addictive.

Before we try to understand the ingestion of "hallucinogens"—smoking dope—we might try to understand smoking. Ordinary smoking of tobacco, as we have seen, turns out to be extraordinary. Only familiarity obscures it.

But, of course, many do "smoke dope," some of it tremendously potent—cocaine, for example.[6] Needless to say, many other "hallucinogenic" substances proliferate, and many other modes of ingesting them. The widespread use of drugs proves that people will not remain penned up in suffocating material comforts—or discomforts. Horizons will be touched and crossed, even if only at random, and with no idea of how to incorporate into the everyday workings of life what is found beyond them. No wonder that lives of novices untutored in the **spiritual life often unravel after their hallucinogenic "trips."

As the body-self cycles and exchanges with the environment, truth is the fruitful building out of the past into present and future. If what is incorporated in "hallucinogenic" experience cannot be woven into the life that follows, we literally do not know what to make of it. To be sure, it sometimes can be made part of life. If not deranged, someone might say and mean, "I have touched for a moment a marvelous realm of power and grace that sustains me. I need not know what it is beyond knowing its power."

Visionary Experiences

Despite dangers of escapism, "hallucinogenic" experiences are serious matters. The novels of Carlos Castaneda, for example, centering around the shamanic use of peyote, reveal more than an urge for diversion. Take the scene in *Tales of Power* in which the shaman leads the novice, after preparation, into a power place in the desert. The younger man experiences an irruptive and dazzling perception of a weird, incandescent beast. But as the experience fades, the novice walks over to the spot and sees nothing but a bush rattling slightly in the wind. He is dismayed. The shaman points out the reason for his disappointment: he is supposing that because it is now only a bush, it could have been nothing else previously.

The novice has absorbed the supposition ubiquitous in mainstream Western civilization and science for millennia: Nature makes no leaps. Things are "substances" that are what they are because they remain identical with themselves, at least over some stretch of time typical for that sort of thing. Change results only through the lawful causal action of the assembly of substances.

But if no such lawful action is detectable, why not assume that a leap from alternate to ordinary reality—an incursion—has occurred? To be sane, we must usually assume the continuity of matter, that Nature does not make leaps. But sometimes it does. This happens on the microscopic level of quanta of energy, and also on the macroscopic with sudden religious conversions. But perhaps more leaps—and *other* leaps—occur than we today typically imagine. Perhaps believing that all ecstasies are computable blocks the greatest ones?[7]

Even if we cannot believe that at that spot in space a weird incandescent beast suddenly was replaced by a bush, the novice's experience jolts us out of the ruts of banal, addictive modes of thought. At the very least it reveals the power of body-self to generate psychic marvels, to connect with unsuspected sources of presence and power.

The Journey

In the dull light of everyday routine, of the to-be-expected, the unexpectable is occluded. But in certain deep moods we realize that there probably are sorts of things going on—probably now in our very midst—that are so strange that we can't imagine them, can't imagine how they would be experienced if we could experience them, so, of course, can't even begin to look for them. They are experienced as beyond our ability to imagine them, that is, as mysterious. In these moods the world is presented in its marvelous opacity, amplitude, and sufficiency. Buddhists ask how such a world could be gathered in our experience. They think it is only an "Emptiness," or a "Void," that could allow this gathering.

Here is an alluring but delicately balanced and risky idea. For, emphatically, it is not a vast emptiness *in* the world—or of course *in* us—

that demands to be filled, but can't be; that is addiction. It is an emptiness that allows the World-whole's presence to appear in its fullness and sufficiency. It is sacramental emptying of ego-self. We began within the womb—and might remain—already full, though we cannot fully comprehend this, nor do we need to.

My wife and I signed-up for a week's introduction to shamanic practices, the course to be conducted by anthropologist Michael Harner. As a young researcher in the jungles of Ecuador, Harner had been initiated with a frighteningly powerful hallucinogenic liquid; he had felt death coming. But in this course no hallucinogens were used, just drums—in fact, shamanic sessions are often called drummings or journeyings.

In easeful trance-like sessions of about twenty minutes, day by day, the shamanic drum resounded flatly, methodically. After sleeping in a tent at night, worried about rain leaking in, I periodically dozed during the first days of the course. Harner's belly laughter would often break through, however, as well as observations like "I can't tell you about metaphysics. Spirits are what you see with your eyes closed."

To warm-up, we would jog in circles around the drummers, seeking a "power animal." The drumbeats loosened up my joints, and I fell into what felt like a horse's body. It was an old horse, but it ran quite well, and I was touched by its stamina, gameness, and love of life. Feeling rejuvenated, I was glad to relate this to others when we described our experiences afterwards.

For the more complex sessions we would lie flat on our backs with scarves over our eyes to block the dim light of the shed in which we gathered. The journeys were always propelled by the steady beating of drums. In the most engrossing sessions, we were asked to image an opening in the Earth down which we would to travel. We were to search for animals to be our allies in solving some difficulty in our lives. "No bugs or spiders," Harner said, but with no laughter.

We went on this journey several times, and the sporadic vividness of the "images" (I did not know what else to call them) was remarkable. But to get down the hole in the Earth required moments of deliberate self-conscious agency, as did the renewal of imagery when the flow of experience animated by the drum no longer held me in its grip.

In the fourth day of the course I again found the old horse, and again was rejuvenated, as if the habitual rhythm of my life had been interrupted by a different one. The new rhythm was no longer experienced as hypnotic and strange, but rather as my proper place and pulse. My earlier habitual routines felt numbing and hypnotic.

But while I felt great, my wife did not. A talented and greatly accomplished

actress and writer, nevertheless her ability to follow Harner's directions for im-
aging things was limited. She could not abandon herself to the flow. Either that
or something else was wrong. The task on this day was to pair up with another
person and to bring back an animal ally to help the partner with some difficulty
in their life. In a voice more plaintive than I ever remembered her using, she
whispered, "Surely you can find some creature to help me. One of them must
know that nobody needs help more than Donnie Esther."

She used the name she was called as a little girl, and her original experience
of powerlessness was in her voice now. The drums began to beat as I lay on the
mat with the scarf over my eyes. For the image of the opening into Earth I had
picked one of the holes at the roots of large creosote bushes in the California
desert. Rodents, sometimes tortoises, had dug them there. This time when I fixed
on the hole it sucked at me and swallowed me into itself. My stomach turned
over, and I feared that being startled would throw me out of the flow. It did not.
I hurtled down a winding shaft, and soon, into the periphery of my vision,
crowded myriad catlike faces, some exceedingly precise and vivid, others incho-
ate. As I asked them for help for Donnie Esther, these presences kept accumu-
lating.

The drums changed to a faster beat and called us back to the surface. We had
been instructed to expel into the body of our partner whatever allies we found,
first into the chest, then into the top of the head. I placed my mouth on her
breastbone and expelled these catlike beings into her chest. So completely did I
empty myself that there were none left to be expelled into her head, although I
went through the motions of doing so. I then whispered into her ear what I had
done.

In the months and years that followed she was accompanied by what she
called "her lioness." It would come and help her when she called it. I noticed
that her behavior changed, and she said she did not feel like an orphan any
more.

There was no need to explain it. There was no need to do anything.

Awe

Reflection, of course, will recuperate: will try to coax concepts to catch
up with intuitions. It might dip back into the archaic sense of truth as
the truth of things or persons—things true to themselves because they
realize their capacities. Out of its past and into its future, body-self is
built out fruitfully. Beyond comprehension, body-self is grafted into the
world. Beyond comprehension there is truth: things work, things work
out.

At the limits of naming, we can only point to the World-whole. I can be—or fail to be—a vital part in an apparently small sector of it, but what my behavior means in the final analysis I cannot tell, because what the universe is, ultimately, I cannot tell. It opens perpetually and surprisingly (and sometimes closes) and presents itself as containing more sorts of things and events than I have ever imagined. *What* it is fills me with awe. More, *that* it is fills me with awe. Though I belong in it, it exists and cycles for no reason I can discover. Any allegedly divine being or any strange event offered as a reason for the World-whole's existence would fall within the Whole, hence leave open the question we cannot escape: Why is there any universe at all? Why not rather nothing?

No answer is forthcoming, but that itself is a sort of answer. I know that nothing I can do will shed more light on the ultimate question. At crucial moments I need do nothing.

Awe undermines addictions.

Awe undermines addictions.

Notes

1. Tom Melham writes of John Muir, "In Alaska, Muir continually sought to experience the great forces of nature; he once climbed a mountain in a blinding snow storm, provoking an Indian companion to remark, 'Muir must be a witch to seek knowledge in such a place . . . and in such miserable weather'." Noting Muir's nearly incredible feats of daring and endurance and, beyond this, the lengths to which he would go to achieve intuitive bonding and insight into Nature, more than one person must have imagined that he tapped certain shamanic powers—though probably Muir himself would have denied this, so strongly did he suspect most claims about "the supernatural" (*John Muir's Wild America*, 13).

2. Only children seem to be candid about their identification with animals. The author of the best-selling *Animorphs* ("They've Been Morphed!") writes, "I get letters from kids talking about what it's like to be a shark or a tarantula. It really captures their imagination" (*USA Today*, Sept. 25, 1997, 8D). But real wild animals capture the imagination more firmly.

3. Lawlor recounts an Aboriginal man's description of a vision: "He said it began by his listening intensely to the sound of a humming bee. He reproduced the sound on the didjereedoo (long wooden flute) so that the bee's body appeared from the flute sound in its Dreaming form. The flute player then dissolved his own body so that it became the humming sound of the bee, thereby entering the bee's body and flying off inside him" (Sources, 382). The Aborigine is aware of what he is doing. Do members of North Atlantic culture have any awareness when they hop into their animal-

machines? Their urges to buy, use, consume can be manipulated to irrational intensities.

4. In Renée Weber, 30–31.

5. In *The Jivaro: People of the Sacred Waterfalls*, Michael Harner writes of a South American tribe who could go into the religious trance simply by standing in a cavern behind their waterfall. As some lost this ability, they began using drugs to achieve the trance.

6. 300 tons of cocaine is consumed in the U.S. per year (NBC evening news, November 21, 1995).

7. All computation, I think, involves computing for a result *within the world*. But ecstatic awe is incited by the very openness of the world *as Whole*. Compare B. L. MacLennan in Pribram, *Rethinking Neural Networks*. Recall references to horizons in our work. Note Roger Penrose, *Shadows of the Mind*, who argues that awareness of felt qualities falls outside of computation, and that there must be a noncomputational level of brain activity. See Dewey on immediately apprehending the quality of whole situations, in his *Logic: The Theory of Inquiry*, New York, 1938, opening chapters.

8. *Desert Solitaire*, 45.

Sources

Abbey, Edward. *Desert Solitaire: A Season in the Wilderness.* New York: Random House-Ballantine, 1977 [1968].

Abram, David. *The Spell of the Sensuous.* New York: Pantheon, 1996.

———."The Ecology of Magic." *Orion: Nature Quarterly,* Summer 1991.

Acampora, Ralph. "Extinction by Exhibition." Connects zoos and pornography, unpublished mss, 1997.

Ackerman, Diane. *A Natural History of the Senses.* New York: Random House-Vintage, 1991 [1990].

Alcoff, Linda, and Elizabeth Porter. *Feminist Epistemologies.* New York: Routledge, 1992.

Alexander, F. M. *The Resurrection of the Body: The Essential Writings of F. Mathias Alexander.* ed. E. Maisel. Including three prefaces to Alexander's books by John Dewey. Boston and London: Shambhala, 1986.

Apffel-Marglin, Frederique. *The Spirit of Regeneration: Andean Culture Confronting Western Notions of Development.* London: Zed Books, 1998.

Appelbaum, David. *Voice.* Albany, NY: SUNY Press, 1990.

Baars, Bernard J. *In the Theatre of Consciousness: The Workspace of the Mind.* New York: Oxford Univ. Press, 1997.

Bachelard, Gaston. *Water and Dreams: An Essay on the Imagination of Matter.* E. Farrell, trans. Dallas: The Institute of Humanities and Culture, 1983 [1942].

Barber, Benjamin. *Strong Democracy: Participatory Politics for a New Age.* Berkeley: Univ. of Calif. Press, 1984.

Barkow, J. H., L. Cosmides, J. Tooby, eds. *The Adapted Mind.* New York: Oxford Univ. Press, 1992.

Barrett, William. *The Illusion of Technique: A Search for Meaning in a Technological Civilization.* Garden City, NY: Doubleday-Anchor, 1978.

———. *Time of Need: Forms of Imagination in the Twentieth Century.* New York:

Harper and Row, 1972. Particularly chapter 4: "Backward Toward the Earth".

Basso, Keith. *Wisdom Sits in Places*. Albuquerque: Univ. of New Mexico Press, 1996.

Bateson, Gregory. *Steps To an Ecology of Mind*. New York: Ballantine, 1972.

Beasley, Conger, Jr. "In Animals We Find Ourselves." *Orion: Nature Quarterly*, Spring 1990.

Beck, Peggy V., et al. *The Sacred: Ways of Knowledge, Sources of Life*. Tsaile, AZ: Navajo Community College Press, redesigned ed., 1992 [1977].

Bell, Diane. *Daughters of the Dreaming*. Melbourne: McPhee Gribble, 1983.

Berman, Morris. *Coming to Our Senses: Body and Spirit in the Hidden History of the West*. New York: Simon and Schuster, 1989.

Berry, Thomas. *The Dream of the Earth*. San Francisco: Sierra Club Books, 1988.

Berry, Thomas, and Brian Swimme. *The Universe Story*. San Francisco: Harper, 1993.

Berry, Wendell. *The Unsettling of America: Culture and Agriculture*. San Francisco: Sierra Club Books, 1977.

———. "Standing by Words," *The Hudson Review*, 33, no. 4, Winter 1980–81.

———. *A Continuous Harmony*. San Diego: Harcourt, Brace, Jovanovich, 1970.

Bevan, Edwyn R. *Hellenism and Christianity*. New York: G. H. Doran, 1922. Especially chapter 8, "Dirt."

Bigwood, Carol. *The Earth Muse: Feminism, Nature, and Art*. Philadelphia: Temple Univ. Press, 1993.

Black Elk. *The Sacred Pipe*. Norman: Univ. of Oklahoma Press, 1953.

Blanchard, Paula. *Margaret Fuller: From Transcendentalism to Revolution*. New York: Dell, 1979.

Bockover, Mary I. *Rules, Rituals, and Responsibility: Essays Dedicated to H. Fingarette*. La Salle, IL: Open Court, 1991.

Bohm, David. *Wholeness and the Implicate Order*. London: Routledge and Kegan Paul, 1980.

———. "Soma-significance: A New Notion of the Relationship Between the Physical and the Mental," ms., 1984, and as indexed in Renée Weber, *Dialogues with Scientists and Sages*. New York: Routledge and Kegan Paul, 1986.

Bordo, Susan. *Unbearable Weight: Feminism, Western Culture, and the Body*. Berkeley: Univ. of Calif. Press, 1993.

Borgmann, Albert. "Artificial Realities: Centering One's Life in an Advanced Technological Setting." In *The Presence of Feeling in Thought*, ed. B. den Ouden and M. Moen, New York: Peter Lang, 1991.

Bowers, C. A. *Critical Essays on Education, Modernity, and the Recovery of the Ecological*. New York: Columbia Univ. Press, 1993.

Buber, Martin. *Between Man and Man*. Boston: Beacon, 1955 [1947].

Burnum Burnum. *Burnam Burnam's Aboriginal Australia: A Traveler's Guide*. Cottage Point, NSW, Australia: Angus and Robertson, 1988.

Burroughs, William S. *Junky*. New York: Penguin, 1977 [1953].

Callicott, J. Baird. *In Defense of the Land Ethic: Essays in Environmental Philosophy*. Albany, NY: SUNY Press, 1989.

Campbell, Joseph. *The Power of Myth: with Bill Moyers*. New York: Doubleday, 1988.

Canetti, Elias. *Crowds and Power*. C. Stewart, trans. New York: Seabury Press, 1978 [1960].

Cannon, Walter B. *The Wisdom of the Body*. New York: W. W. Norton, 1963 [1932, revised and enlarged edition, 1939].

Casey, Edward S. *Getting Back Into Place*. Bloomington and Indianapolis: Indiana Univ. Press, 1993.

————. *Remembering: A Phenomenological Study*. Bloomington and Indianapolis: Indiana Univ. Press, 1987.

Castaneda, Carlos. *Tales of Power*. New York: Pocket Books, 1976.

Cather, Willa. *The Professor's House*. New York: Vintage Books, 1990. A subtle study of a deracinated intellectual.

Chatwin, Bruce. *The Songlines*. New York: Viking-Penguin, 1987.

Chodorow, Nancy. *The Reproduction of Mothering: Psychoanalysis and the Sociology of Gender*. Berkeley: Univ. of Calif. Press, 1978.

Chuang Tzu. *Basic Writings*. B. Watson, trans. New York: Columbia Univ. Press, 1964.

Cobb, John B, Jr. *Process Theology as Political Theology*. Philadelphia: Fortress Press, 1982.

————, and Herman E. Daly. *For the Common Good* (see Daly below).

Cooper, Ron L. *Whitehead and Heidegger: A Phenomenological Investigation into the Intelligibility of Experience*. Athens, Ohio: The Univ. Press, 1993.

Cosmides, Leda, et al. eds. *The Adapted Mind: Evolutionary Psychology and the Generation of Culture*. New York: Oxford Univ. Press, 1992.

Csikszentmihalyi, Mihaly. *The Evolving Self: A Psychology for the Third Millennium*. New York: Harper-Collins, 1993.

Daly, Herman E., and John B. Cobb, Jr. *For the Common Good: Redirecting the Economy Toward Community, the Environment, and a Sustainable Future*. Boston: Beacon Press, 1989.

Damasio, Antonio R. *Descartes' Error: Emotion, Reason, and the Human Brain*. New York: G. P. Putnam's Sons, 1994.

de las Casas, Bartolome. *The Devastation of the Indies—A Brief Account*. B. M.

Donovan, ed. H. Briffault, trans. Baltimore: Johns Hopkins Univ. Press, 1992. A sixteenth century work.

de Waal, Frans. *Good Natured: The Origins of Right and Wrong in Humans and Other Animals*. Cambridge: Harvard Univ. Press, 1996.

Dewey, John. *Art as Experience*. New York: G. P. Putnam's Sons, 1980 [1934].

————. *Experience and Nature*. 2nd ed. New York: Dover, 1958 [1929].

————. *Human Nature and Conduct*. New York: The Modern Library, 1930 [1922].

————. *Logic: The Theory of Inquiry*. New York: Holt, Rinehart, and Winston, 1964 [1938].

————. *The Poems of John Dewey*, ed. Jo Ann Boydston. Carbondale, IL: So. Ill. Univ. Press, 1977.

Dole, Vincent. "An Apparent Unwisdom of the Body." Foreword to *Drug Abuse: Clinical and Basic Concepts*. N. Pradhan and N. Dutta, eds. St. Louis: Mosby, 1977.

Dostoevsky, Fyodor. *The Gambler: With the Diary of Polina Suslova*. V. Terras, trans. Chicago: The Univ. Press, 1972.

Douglas, Mary. *Purity and Danger: An Analysis of the Concepts of Pollution and Taboo*. London: Routledge and Kegan Paul, 1966.

Doyle, David E., ed. *Anasazi Regional Organization and the Chaco System*. Albuquerque: Univ. of New Mexico, 1992. (Maxwell Museum of Anthropology, Anthropological Papers No. 5). Particularly "Anasazi Ritual Landscapes," J. R. Stein and S. H. Lekson.

Driver, B. L, et al. *Nature and the Human Spirit: Toward an Expanded Land Management Ethic*. State College, PA: Venture Publications, 1996.

Duerr, Hans Peter. *Dreamtime: Concerning the Boundary Between Wilderness and Civilization*. F. Goodman, trans. New York: Basil Blackwell Inc., 1985 [1978].

Dustin, Daniel. *The Wilderness Within*. San Diego: S. D. State Univ. Institute for Leisure Behavior, 1993.

Edwards, Jonathan. *Devotions of Jonathan Edwards*. Grand Rapids: Baker Book House, 1959.

Ehrenfeld, David. *The Arrogance of Humanism*. New York: Oxford Univ. Press, 1978.

————. *Beginning Again: People and Nature in the New Millennium*. New York: Oxford Univ. Press, 1993. Includes crucial articles: "Changing the Way We Farm," "Forgetting," "A New Role for Experts," "Life in the New Millennium."

Eliade, Mircea. *The Myth of the Eternal Return, or, Cosmos and History*. W. Trask, trans. Princeton: The Univ. Press, 1974 [1949].

Emerson, R. W. *Ralph Waldo Emerson: Selected Essays*, ed. L. Ziff. New York:

Penguin, 1982. Encompassing nearly all of the citations from Emerson in the present volume. Our opening epigraph is from the last lines of "Circles."

Erikson, Eric. *The Life Cycle Completed*. New York: W. W. Norton, 1982.

Erikson, Joan. "Eye to Eye," in *The Man-Made Object*. G. Kepes, ed. New York: G. Braziller, 1966.

Evans, Arthur. *Critique of Patriarchal Reason*. San Francisco: White Crane Press, 1997.

Faulkner, Rozanne W. *Therapeutic Recreation Protocol for Treatment of Substance Addictions*. State College, PA: Venture Publishing, 1991.

Feher, Michel, et al. eds. *Fragments for a History of the Human Body*. 3 vols. New York: Zone Publishers, 1989.

Feldenkrais, Moishe. *Body and Mature Behavior: A Study of Anxiety, Sex, Gravitation and Learning*. New York: International Universities Press, 1979 [1949].

Fingarette, Herbert. *Heavy Drinking*. Berkeley: Univ. of Calif. Press, 1988.

Fox, Mathew. "Addiction in Overdeveloped Cultures." *Creation Spirituality*, vol. 7, no. 1, Oakland, CA, 1991.

Fromm, Eric. *Escape from Freedom*. New York: Farrar and Rinehart, 1941.

Gadon, Elinor W. *The Once and Future Goddess: A Symbol for Our Time*. San Francisco: Harper, 1989.

Galanter, Marc and Herbert D. Kleber. *Textbook of Substance Abuse Treatment*. Washington DC: American Psychiatric Press, 1994.

Gardner, Kay. *Sounding the Inner Landscape: Music as Medicine*. Stonington, ME: Caduceus Publications, 1990.

Gebser, Jean. *The Ever Present Origin*. N. Barstad and A. Mikunas, trans. Athens, Ohio: The Univ. Press, 1985 [1949, 1953].

Gendlin, Eugene T. "Thinking Beyond Patterns: Body, Language, and Situations." In *The Presence of Feeling in Thought*, ed. by B. den Ouden and M. Moen. New York: Peter Lang, 1991.

Gilligan, Carol. *In a Different Voice*. Cambridge: Harvard Univ. Press, 1982.

Gimbutas, Marija. *Civilization of the Goddess: The World of Old Europe*. San Francisco: Harpers, 1991.

———. *The Language of the Goddess*. San Francisco: Harper, 1989.

———. *The Goddesses and Gods of Old Europe: 6500–3500 B.C.: Myths and Occult Images*. Berkeley: Univ. of Calif. Press, 1982.

Glatt, M. M. *A Guide to Addiction and Its Treatment*. New York: Wiley, 1974.

Glendinning, Chellis. *My Name is Chellis, and I'm in Recovery from Western Civilization*. Boston: Shambhala, 1994.

———. "Dreaming for the Earth." *Woman of Power*, Spring 1991.

Goldenberg, Naomi. *Changing of the Gods: Feminism and the End of Traditional Religions*. Boston: Beacon Press, 1979.

Goldfarb, William. "Groundwater: The Buried Life." In *Ecology, Economics, Ethics: Closing the Broken Circle*, ed. by Bormann and Kellert. New Haven: Yale Univ. Press, 1992. An exemplary use of both mythic and scientific approaches to ecological challenges.

Good, Kenneth (with David Chanoff). *Into the Heart: One Man's Pursuit of Love and Knowledge Among the Yanomama*. London: Penguin, 1992.

Goodman, Felicitas D. *Ecstasy, Ritual, and Alternate Reality: Religion in a Pluralistic World*. Bloomington and Indianapolis: Indiana Univ. Press, 1988.

————. *How About Demons? Possession and Exorcism in the Modern World*. Bloomington and London: Indiana Univ. Press, 1988.

Gore, Al. *Earth in the Balance: Ecology and the Human Spirit*. New York: Penguin, 1993.

Gorski, Terence T. *Understanding the Twelve Steps*. New York: Prentice Hall/Parkside, 1989.

Greaves, Thomas. *Intellectual Property Rights for Indigenous Peoples: A Source Book*. Oklahoma City: Society for Applied Anthropology, 1994.

Griffin, David Ray. *Sacred Interconnections: Postmodern Spirituality, Political Economy, and Art*. Albany: SUNY Press, 1990.

Griffin, Susan. *Woman and Nature: The Roaring Inside Her*. New York: Harper and Row, 1978.

Grim, John A. *The Shaman: Patterns of Religious Healing Among the Ojibway Indians*. Norman, OK: The Univ. Press, 1982.

Grosz, Elizabeth. *Volatile Bodies: Toward a Corporeal Feminism*. Bloomington and Indianapolis: Indiana Univ. Press, 1994.

Hall, Nor, and Warren R. Dawson. *Broodmales: A Psychological Essay on Men in Childbirth*. Dallas: Spring Publications, 1989. Dawson's essay, here extensively introduced by Hall, first appeared in 1929.

Hanke, Lewis. *Aristotle and the American Indians*. Bloomington: Indiana Univ. Press, 1959. Traces deeply rooted bias against native peoples.

Haraway, Donna J. *Simians, Cyborgs, and Women: The Reinvention of Nature*. London: Free Association, 1991.

Harner, Michael. *The Way of the Shaman*. San Francisco: Harper and Row, 1980.

Hartig, T., et al. "Perspectives on Wilderness: Testing the Theory of Restorative Environments." In T. Easley and J. Passineau, *The Use of Wilderness for Personal Growth*. U.S. Dept. of Agriculture, Supt. of Documents, #13.88; Rm-193, 1990.

Hawken, Paul. *The Ecology of Commerce*. New York: Harper Business, 1993.

Heidegger, Martin. *Discourse on Thinking*. J. Anderson and and E. H. Freund, trans. New York: Harper and Row, 1966.

Heim, Michael. *The Metaphysics of Virtual Reality*. New York: Oxford Univ. Press, 1993.

Heller, Joseph, and William Henkin. *Bodywise: Regaining Your Natural Flexibility and Vitality for Maximum Well-Being*. Los Angeles: J. P. Tarcher, 1986.

Hickman, Larry A. *John Dewey's Pragmatic Technology*. Bloomington: Indiana Univ. Press, 1990.

———, ed., *Reading Dewey: Interpretations for a Postmodern Generation*. Bloomington: Indiana Univ. Press, 1998.

Highwater, Jamake. *The Primal Mind: Vision and Reality in Indian America*. New York: New American Library, 1981.

Hillman, James. *The Soul's Code*. New York: Random House, 1996.

———. *Revisioning Psychology*. New York: Harper and Row, 1977.

Hocking, William Ernest. *The Self: Its Body and Freedom*. New Haven: Yale Univ. Press, 1928.

Hopkins, Gerard Manley. *Poems of Gerard Manley Hopkins*. Mt. Vernon, NY: Peter Pauper Press, no date.

House, Freeman. "Totem Salmon." Obtainable from Way of the Mountain Learning Center, 585 E. 31st St., Durango, CO, 81301.

Irigary, Luce. *I Love to You*. A. Martin, trans. New York: Routledge, 1996.

———. *Marine Lover of Friedrich Nietzsche*. G. Gill, trans. New York: Columbia Univ. Press, 1991.

———. *This Sex Which Is Not One*. C. Porter, trans. Ithaca, NY: Cornell Univ. Press, 1985.

Jackson, Michael. *Paths Toward a Clearing: Radical Empiricism and Ethnographic Inquiry*. Bloomington and Indianapolis: Indiana Univ. Press, 1989.

Jacobsen, Thorkild. *The Treasures of Darkness: History of Mesopotamian Religion*. New Haven: Yale Univ. Press, 1976.

Jaggar, Alison M. *Feminist Politics and Human Nature*. Totowa, NJ: Rowman and Allanheld, 1983.

James, William. *Essays in Radical Empiricism* and *A Pluralistic Universe*. New York: Longmans Green and Co., 1958 [1912].

———. *The Varieties of Religious Experience*. New York: New American Library, 1958 [1902].

———. *The Principles of Psychology*. New York: Dover Books, 1950 [1890].

———. *William James: The Essential Writings*, ed. B. Wilshire. Albany: SUNY Press, 1984 [1971]. Selections from, e.g., *The Principles of Psychology*, *The Varieties of Religious Experience*, "The Place of Affectional Facts in a World of Pure Experience," "The Will to Believe," all of "Humanism and Truth" and "Does Consciousness Exist?"

Jaspers, Karl. *The Idea of the University*. H. Reiche and H. Vanderschmidt, trans. Boston: Beacon Press, 1959 [1946].

Ji, Sungchul. "Complementarism: A Biology-Based Philosophical Framework to Integrate Western Science and Eastern Tao." *Proceedings: 16th International Congress of Psychotherapy*, 1994, Korean Academy of Psychotherapists.

Johnson, Buffie. *The Lady of the Beasts*. San Francisco: Harper, 1989.

Johnson, Mark. *The Body in the Mind: The Bodily Basis of Meaning, Imagination, and Reason*. Chicago: The Univ. Press, 1987.

Jones, Frank P. *The Alexander Technique: Body Awareness in Action*. New York: Schocken Books, 1976. Particularly the chapter on Dewey and Alexander.

Jung, Carl, and Carl Kerenyi. *Essays on a Science of Mythology*. Princeton: The Univ. Press, 1969.

Kaplan, Rachel and Stephen Kaplan. *The Experience of Nature: A Psychological Perspective*. Cambridge (Eng.) and New York: Cambridge Univ. Press, 1989.

Kaplan, S., and J. Talbot. "Psychological Benefits of Wilderness Experience." In I. Altman and J. Wohlwill, eds. *Behavior and Natural Environment*. New York: Plenum Press, 1983.

Kasl, Charlotte Davis. *Women, Sex, and Addiction: A Search for Love and Power*. New York: Harper and Row, 1989.

Keller, Catherine. *From a Broken Web: Separation, Sexism, and Self*. Boston: Beacon Press, 1986. Importantly interconnects Whiteheadian and feminist viewpoints.

Kellert, Stephen, and E. O. Wilson, eds. *The Biophilia Hypothesis*, Washington D.C.: Island Press, 1993.

Kemeny, Margaret. "Emotions and the Immune System," in Moyers, *Healing and the Mind*. New York: Doubleday, 1993.

Kestenbaum, Victor. *The Phenomenological Sense of John Dewey*. New York: Humanities Press, 1978.

Kierkegaard, Soren. *The Concept of Dread*. W. Lowrie, trans. Princeton: The Univ. Press, 1957 [1844].

Klein, Bob. *Movements of Magic: The Spirit of T'ai-Chi-Ch'uan*. N. Hollywood, CA: Newcastle Pub. Co., 1984.

Knight, Chris. *Blood Relations: Menstruation and the Origins of Culture*. New Haven: Yale Univ. Press, 1991.

Kolakowski, Leszek. *The Presence of Myth*. Chicago: The Univ. Press, 1988 [1972].

Korten, David. *When Corporations Rule the World*. San Francisco: Berrett-Koehler Publishers, 1996.

Krishnamurti, Jiddu. *Freedom from the Known*. San Francisco: Harper, 1969.

Kristeva, Julia. *The Kristeva Reader*. T. Moi, ed. New York: Columbia Univ. Press, 1986.

La Barre, Weston. *The Peyote Cult*. New York: Shoestring Press, 1970 [1959].

———. *The Ghost Dance: The Origins of Religion*. Garden City, NY: Doubleday, 1970.

LaChapelle, Dolores. *Sacred Land Sacred Sex—Rapture of the Deep: Concerning Deep Ecology and Celebrating Life*. Silverton, CO.: Finn Hill Arts, 1988.

Lachs, John. *The Relevance of Philosophy to Life*. Nashville: Vanderbilt Univ. Press, 1995.

Lanier, Jaron. "Virtual Reality." A talk given to The Reality Club, New York City, February 28, 1990.

Lao Tzu. *Tao Te Ching*. D. C. Lau, trans. Hong Kong: Chinese University Press, 1963 (in English and Chinese). The quoted passage is para. 25.

Lauter, Estella, and Carol Schreier Rupprecht. *Feminist Archetypal Theory: Interdisciplinary Re-Visions of Jungian Thought*. Knoxville: Univ. of Tenn. Press, 1985.

Lawlor, Robert. *Voices of the First Day: Awakening in the Aboriginal Dreamtime*. Rochester, VT: Inner Traditions, 1991.

Levinthal, Charles F. *Messengers of Paradise: Opiates and the Brain—The Struggle over Pain, Rage, Uncertainty and Addiction*. Garden City, NY: Doubleday/Anchor, 1988.

Lévi-Strauss, Claude. *Structural Anthropology*, especially chap. 10, "The Effectiveness of Symbols." C. Jacobson and B. Schoepf, trans. New York: Basic Books, 1963.

Lévy-Bruhl, Lucien. *The Notebooks on Primitive Mentality*. P. Rivère, trans. New York: Harper, 1978.

Lifton, Robert Jay. *The Nazi Doctors: Medical Killing and the Psychology of Genocide*. New York: Basic Books, 1986.

Lincoln, Bruce. *Emerging From the Chrysalis: Studies in Rituals of Women's Initiation*. Cambridge, MA: Harvard Univ. Press, 1981.

Lingis, Alphonso. *The Community of Those Who Have Nothing in Common*. Bloomington: Indiana Univ. Press, 1994.

Lizot, Jacques. *Tales of the Yanomami*. Cambridge (Eng.) and New York: Cambridge Univ. Press, 1985 [1976].

Lovelock, James. *Gaia: A New Look at Life on Earth*. New York: Oxford Univ. Press, 1987.

Lowinson, Joyce H., et al. *Substance Abuse: A Comprehensive Textbook*. 2nd ed. Baltimore: Williams and Wilkins, 1992.

Mahdi, Louise Carus, et al. *Betwixt and Between: Patterns of Masculine and Feminine Initiation*. La Salle, IL: Open Court, 1987.

Mang, M. *The Restorative Effects of Wilderness Backpacking*. Unpublished dissertation. Univ. of California, Irvine, 1984.

Mannell, Roger. "Approaches in the Social and Behavioral Science to the Systematic Study of Hard-to-Define Human Values and Experiences." In *Nature and the Human Spirit*, ed. B. L. Driver, et al. State College, PA: Venture Publishing, 1996.

Marshack, Alexander. *The Roots of Civilization: The Cognitive Beginnings of Man's First Art, Symbol and Notation*. New York: McGraw-Hill, 1972.

Martin, Calvin Luther. *The Way of the Human Being*. New Haven: Yale Univ. Press, 1998.

———. *In the Spirit of the Earth: Rethinking History and Time*. Baltimore and London: Johns Hopkins Univ. Press, 1992.

Maser, Chris. *The Redesigned Forest*. San Pedro, CA: R & E Miles, 1988.

Maslow, Abraham H. *Motivation and Personality*. 3d ed. New York: Harper and Row, 1987 [1954].

Mazis, Glen A. *The Trickster, Magician, and Grieving Man: Reconnecting Men with Earth*. Santa Fe, NM: Bear and Co., 1994.

———. *Emotion and Embodiment: Fragile Ontology*. New York: Peter Lang, 1993.

McKibben, Bill. *The End of Nature*. New York: Random House, 1989.

Merchant, Caroline. *The Death of Nature*. San Francisco: Harper and Row, 1980.

Merleau-Ponty, Maurice. *The Visible and the Invisible* (followed by working notes). A. Lingis, trans. Evanston, IL: Northwestern Univ. Press, 1968.

Midgley, Mary. *Beast and Man: The Roots of Human Nature*. Ithaca, NY: Cornell Univ. Press, 1978.

Milhorn, Howard T. *Chemical Dependence: Diagnosis, Treatment, and Prevention*. New York: Springer-Verlag, 1990.

Mithen, Steven. *The Prehistory of the Mind: The Cognitive Origins of Art, Religion, and Science*. New York: Thames and Hudson, 1996.

Montague, Margaret Prescott. "Twenty Minutes of Reality." First published anonymously in the *Atlantic Monthly* in 1916, and in pamphlet form, 1947, by the Macalister Park Pub. Co., St. Paul, MN.

Montejo, Victor. *Testimony: Death of a Guatemalan Village*. V. Perera, trans. Willamatic, CT: Curbstone Press, 1987.

Muir, John. *My First Summer in the Sierra*, preface by F. Turner. San Francisco: Sierra Club, 1988 [1911]. Condensed in *Gentle Wilderness: The Sierra Nevada*, ed. D. Brower, brilliant photos by R. Kauffman, San Francisco: Promontory Press, 1981 [1967].

———. *John of the Mountains: The Unpublished Journals of John Muir*, ed. L. M. Wolfe. Madison: Univ. of Wisconsin Press, 1938.

Muller, E. E., and A. R. Genazzani. *Central and Peripheral Endorphins*. New York: Raven Press, 1984.

Nabokov, Peter. *Native American Testimony: A Chronicle of Indian-White Relations from Prophecy to the Present, 1492–1992*. New York: Penguin, 1992 [1991].

Naess, Arne. "*Self* Realization: An Ecological Approach to Being in the World." In *Thinking Like a Mountain*, ed. J. Seed. Santa Cruz: New Society Publishers, 1988.

Nash, Roderick. *Wilderness and the American Mind*. New Haven: Yale Univ. Press, 1982.

Neihardt, John G. *Black Elk Speaks*. Lincoln: Univ. of Nebraska Press, 1979 [1932].

Neville, Robert, *The High Road Around Modernism*. Albany, NY: SUNY Press, 1992.

Nietzsche, Friedrich. *Thus Spake Zarathustra*. W. Kaufmann, trans. New York: Viking, 1978. Particularly "The Despisers of the Body."

Northrup, Christiane. *Womens' Bodies—Womens' Wisdom*. New York: Bantam, 1994.

Nye, Andrea. *Words of Power: A Feminist Reading of the History of Logic*. New York: Routledge, 1990.

O'Connell, Robert L. *Ride of the Second Horseman: The Birth and Death of War*. New York: Oxford Univ. Press, 1995.

Ogilvy, James, ed. *Revisioning Philosophy*. Albany, NY: SUNY Press, 1992.

Olds, David D. "Consciousness: A Brain Centered, Informational Approach." *Psychoanalytic Inquiry*, vol. 12, 1992, 419–444.

Orr, David W. *Earth in Mind: On Education, Environment, and the Human Prospect*. Washington, DC: Island Press, 1994.

———. *Ecological Literacy: Education and the Transition to the Postmodern World*. Albany, NY: SUNY Press, 1991.

Osborne, Lawrence. *The Poisoned Embrace: A Brief History of Sexual Pessimism*. New York: Pantheon, 1993.

Pauli, Wolfgang. *Writings on Physics and Philosophy*. R. Schlapp, trans. Berlin and NY: Springer-Verlag, 1994.

Pearce, Joseph Chilton. *Magical Child: Rediscovering Nature's Plan for Our Children*. New York: Dutton, 1977.

Penrose, Roger. *Shadows of the Mind: A Search for the Missing Science of Consciousness*. New York: Oxford Univ. Press, 1994.

Perkins, John. *Shape Shifting: Shamanic Techniques for Global and Personal Transformation*. Rochester, VT: Destiny Books, 1997.

Pert, Candace. "The Chemical Communicators." In Moyers, *Healing and the Mind*. New York: Doubleday, 1993.

Pettit, Jan. *Utes: The Mountain People*. Boulder, CO: Johnson Books, 1990.

Piaget, Jean, and B. Inhelder. *The Growth of Logical Thinking From Childhood to*

Adolescence: An Essay in the Construction of Formal Operational Structures. New York: Basic Books, 1958.

Plumwood, Val. "Nature, Self, and Gender: Feminism, Environmental Philosophy, and Critique of Rationalism." In *Environmental Philosophy*, ed. Michael Zimmerman, 1992.

Poirier, Richard. *The Renewal of Literature: Emersonian Reflections*. New Haven: Yale Univ. Press, 1988.

Pratt, Annis. *Dancing With Goddesses: Archetypes, Poetry, and Empowerment*. Bloomington: Indiana Univ. Press, 1994.

Pribram, Karl H. *Brain and Perception: Holonomy and Structure in Figural Processing*. Hillsdale NJ, Hove and London: Lawrence Earlbaum Associates, 1991.

———, ed. *Origins: Brain and Self-Organization*, Hillsdale NJ, Hove and London: Lawrence Earlbaum Associates, 1994.

———, ed. *Rethinking Neural Networks*. Hillsdale NJ, Hove and London: Lawrence Earlbaum Associates, 1993.

Prigogine, Ilya. "Mind and Matter: Beyond the Cartesian Dualism." In Pribram, 1994, above.

Prigogine, Ilya, and Isabelle Stengers. *Order Out of Chaos: Man's Dialogue with Nature*. New York: Bantam, 1984.

Putnam, Hilary. *Renewing Philosophy*. Cambridge: Harvard Univ. Press, 1992.

Redner, Harry. *A New Science of Representation: Towards an Integrated Theory of Representation in Science, Politics, and Art*. Boulder, CO: Westview Press, 1994.

Rice, Julian. *Black Elk's Story: Distinguishing Its Lakota Purpose*. Albuquerque: Univ. of New Mexico Press, 1991.

Robinson, B. E. *Work Addiction: Hidden Legacies of Adult Children*. Deerfield Beach, FL: Health Communications, 1989.

Rockefellar, Steven C. *Religious Faith and Democratic Humanism: An Essay in the Life and Thought of John Dewey*, New York: Columbia Univ. Press, 1991.

Romanyshyn, Robert. "Dream Body in Cyberspace." Internet, 02/07/96 11: 56: 31 http: www.eden.com/magical /45/dream_body.html.

Ross, Colin A. *Multiple Personality Disorder: Diagnosis, Clinical Features, and Diagnosis*. New York: Wiley Interscience, 1989.

Ruden, Ronald. *The Craving Brain: The Biobalance Approach to Controlling Addictions*. New York: Harper-Collins, 1997.

Ryan, Paul. *Video Mind, Earth Mind: Art, Communications, and Ecology*. New York: Peter Lang, 1992. Important collected articles.

Sahlins, Marshall. *Stone Age Economics*. Chicago: The Univ. Press, 1972.

Sanchez, Carol Lee. "New World Tribal Communities: An Alternative Approach for Recreating Egalitarian Societies." In J. Plaskow and C. Christ, eds. *Weaving the Visions*, San Francisco: Harper and Row, 1989.

Sarles, Harvey B. *Language and Human Nature*. Minneapolis: Univ. of Minn. Press, 1985 [1977].

Sauer, Carl. "Theme of Plant and Animal Destruction in Economic History." *Journal of Farm Economics*, 20, 1938: 765–775; cited in *Sacred Land, Sacred Sex—Rapture of the Deep*. Dolores La Chappele. Silverton, CO, 1988, p. 36.

Schaef, Anne Wilson. *When Society Becomes an Addict*. San Francisco: Harper and Row, 1987.

Schilder, Paul. *The Image and Appearance of the Human Body*. New York: International Universities Press, 1950.

Schroeder, Herbert. "Psyche, Nature, and Mystery: Some Psychological Perspectives on the Values of Natural Environments." In Driver, et al. *Nature and the Human Spirit*. State College, PA: Venture Publishing, 1996.

Schuckit, Marc A. *Drug and Alcohol Abuse: A Clinical Guide to Diagnosis and Treatment*. 4th ed. New York: Plenum Medical Book Co., 1995.

Schumacher, John A. *Human Posture: The Nature of Inquiry*. Albany, NY: SUNY Press, 1989. Unheralded but essential.

Searles, Harold. *The Nonhuman Environment—In Normal Development and in Schizophrenia*. New York: International Universities Press, 1960.

Selye, Hans. *The Stress of Life*. Rev. ed. New York: McGraw-Hill, 1978 [1956].

———. *Stress Without Distress*. New York: New American Library, 1975 [1974].

Sered, Susan Starr. *Priestess, Mother, Sacred Sister: Religions Dominated by Women*. New York: Oxford Univ. Press, 1994.

Sessions, George, *Ecophilosophy*. Philosophy Department, Sierra College, Rocklin, CA, 95677. An ongoing publication since 1976.

Shaffii, Mohammad, and Sharon Lee Shaffii, eds. *Biological Rhythms, Mood Disorders, Light Therapy, and the Pineal Gland*. Washington, DC: American Psychiatric Press, 1990.

Sheets-Johnstone, Maxine. *The Roots of Power: Animate Form and Gendered Bodies*. Peru, IL: Open Court, 1994.

———. *The Roots of Thinking*. Philadelphia: Temple Univ. Press, 1990.

Sheldrake, Rupert. *Seven Experiments that Could Change the World*. New York: Riverhead Books, 1995.

Shepard, Paul. *The Others: How Animals Made Us Human*. Washington, DC and Covelo, CA: Island Press, 1996.

———. *Nature and Madness*. San Francisco: Sierra Club Books, 1982.

Sikorsky, Igor I., Jr. *AA's Godparents: Carl Jung, Emmet Fox, Jack Alexander: Three Early Influences on Alcoholics Anonymous and Its Foundation*. Minneapolis: CompCare Publishers, 1990.

Silko, Leslie Marmon. *Ceremony*. New York: Quality Paperback Book Club, 1994 [1977].

Skaggs, Merrill Maguire. *After the World Broke in Two: The Later Novels of Willa Cather*. Charlottesville and London: Univ. of Virginia Press, 1990.

Smith, Quentin. *The Felt Meanings of the World: A Metaphysics of Feeling*. W. Lafayette, IN: Purdue Univ. Press, 1986. A nearly unheralded but basic book.

Snyder, Gary. *The Practice of the Wild*. San Francisco: North Point Press, 1990.

———. *The Old Ways*. San Francisco: City Lights Books, 1977.

Somé, Malidoma Patrice. *Of Water and Spirit: Ritual, Life, and Initiation in the Life of an African Shaman*. New York: Penguin, 1995.

Stich, Stephen. *The Fragmentation of Reason*. Cambridge: MIT Press, 1990.

Straus, Erwin. "The Expression of Thinking." In *An Invitation to Phenomenology*, ed. J. Edie. Chicago: Quadrangle Books, 1965.

Stuhr, John, ed. *Philosophy and the Reconstruction of Culture: Essays After Dewey*. Albany, NY: SUNY Press, 1993.

Sullivan, Lawrence E. "The Attributes and Power of the Shaman: A General Description of the Ecstatic Care of the Soul." In *Ancient Traditions: Shamanism in Central Asia and the Americas*. G. Seaman and J. Day, eds. Denver: Univ. of CO and Denver Museum of Natural History, 1994.

Teilhard de Chardin, Pierre. *Building the Earth and The Psychological Conditions of Human Unification*. N. Lindsay and J. Caulfield, trans. New York: Discus Books, 1969.

Terman, Michael. "Light Treatment." In *Principles and Practice of Sleep Medicine*, M. Kryger et al. eds. 2nd ed. Philadelphia: Saunders, 1994.

Thompson, Becky W. *A Hunger So Wide and So Deep: American Women Speak Out on Eating Problems*. Minneapolis and London: Univ. of Minnesota Press, 1994.

Thompson, William Irwin. *The Time Falling Bodies Take to Light: Mythology, Sexuality, and the Origins of Culture*. New York: St. Martin's Press, 1981.

Thoreau, Henry David. *Walden: Or, Life in the Woods*. New York: New American Library, 1960.

———. "Walking." San Francisco: Harper, 1994.

Tiger, Lionel. *Optimism: The Biology of Hope*. New York: Simon and Schuster, 1979. Especially chapter 4.

Todd, Nancy Jack, and John Todd. *Bioshelters, Ocean Arks, City Farming: Ecology as the Basis of Design*. San Francisco: Sierra Club Books, 1984.

Todes, Samuel. *The Body as the Material Subject of the World*. Harvard Dissertations in Philosophy. New York: Garland Publishing Co., 1989. A classic, published twenty-six years after it was written.

Tooby, J., and L. Cosmides, "The Psychological Foundations of Culture." In *The Adapted Mind*, ed. by Tooby, Cosmides, and Barkow. New York: Oxford Univ. Press, 1992.

Tucker, Mary Evelyn and D. R. Williams, eds. *Buddhism and Ecology: The Interconnection of Dharma and Deeds*. Cambridge: Harvard Univ. Press, 1997. Note Lawrence Sullivan's preface.

———. *The Ecological Spirituality of Teilhard*. Chambersburg, PA: Teilhard Studies, Spring 1985.

——— with John Grim, eds. *Worldviews and Ecology*. Lewisburg, PA: Bucknell Univ. Press, 1993.

Turner, Frederick W. *Beyond Geography: The Western Spirit Against Wilderness*. New Brunswick, NJ: Rutgers Univ. Press, 1990 [1983].

Turner, Victor "Betwixt and Between: The Liminal Period in Rites of Passage." In *Betwixt and Between*, ed. L. C. Mahdi. La Salle, IL: Open Court, 1987.

Van Dresser, Peter. "The Coming Solar Age" and "The Modern Retreat from Function." In *The Subversive Science: Essays Toward an Ecology of Man*. P. Shephard and D. McKinley, eds. Boston: Houghton Mifflin, 1969.

Weber, R. J., and C. B. Pert. "Opiatergic Modulation of the Immune System." In Muller and Genazzani, eds. *Central and Peripheral Endorphins*. New York: Raven Press, 1984.

Weil, Andrew. *The Natural Mind: A New Way of Looking at Drugs and the Higher Consciousness*. Boston: Houghton Mifflin, 1972.

——— and Winifred Rosen. *From Chocolate to Morphine: Everything You Need to Know about Mind Altering Drugs*. Boston: Houghton Mifflin, 1983.

West, Cornel. *The American Evasion of Philosophy: A Genealogy of Pragmatism*. Madison, Wisconsin: The Univ. Press, 1989.

Whitehead, Alfred North. *Process and Reality: An Essay in Cosmology*. Corrected edition. New York: Free Press, 1978 [1929].

Williamson, Ray A. *Living the Sky: The Cosmos of the American Indian*. Norman, OK: The Univ. Press, 1984.

Wilshire, Bruce. *The Moral Collapse of the University: Professionalism, Purity, and Alienation*. Albany, NY: SUNY Press, 1990.

———. *Role Playing and Identity: The Limits of Theatre as Metaphor*. Bloomington: Indiana Univ. Press, 1991 [1982].

———. *William James and Phenomenology: A Study of 'The Principles of Psychology.'* Bloomington: Indiana Univ. Press, 1968. AMS edition, NY, 1979.

———, ed. *The Essential Writings of William James*. Albany, NY: SUNY Press, 1984 [1971]. See above under James.

———. "The Breathtaking Intimacy of the Material World: William James' Last Thoughts." In *Companion to William James*. ed. R. A. Putnam. New York: Cambridge Univ. Press, 1997.

———. "Edie's Hard Nosed James and the Retrieval of the Sacred." In *James*

Edie: Phenomenology and Scepticism, B. Wachterhauser, ed. Evanston, IL: Northwestern Univ. Press, 1996.

———. "John Dewey's Views on the Subconscious: Difficulties in the Reconstruction of Culture." In *Philosophy and the Reconstruction of Culture*, ed. J. Stuhr. Albany, NY: SUNY Press, 1993.

———. "Mimetic Engulfment and Self-Deception." In *Perspectives on Self-Deception*. A. Rorty and B. MacLaughlin, eds. Berkeley: The Univ. of Calif. Press, 1989.

Wilshire, Donna. *Virgin, Mother, Crone: Myths and Mysteries of the Triple Goddess*. Rochester VT: Inner Traditions, 1994.

———. *Virgin, Mother, Crone*. Audio-casette enactment. Inner Traditions, Rochester, VT, 1995.

———. "The Uses of Image, Myth, and the Female Body in Re-Visioning Knowledge." In *Gender, Body, Knowledge: Feminist Reconstructions of Knowing and Being*, ed. by A. Jaggar and S. Bordo. New Brunswick, NJ: Rutgers Univ. Press, 1989.

Wilshire, Leland E. "The Servant City: A New Interpretation of the 'Servant of the Lord' in the Servant Songs of Deutero-Isaiah," *Journal of Biblical Literature*, 94, 1975.

———. "Jerusalem as the 'Servant City' . . . Reflections in the Light of Further Study . . ." In *The Bible in the Light of Cuneiform Literature*. Lewiston, ME: Edwin Mellon, 1990. Studies of corporate individuation in a Biblical context.

Wilson, E. O. *Biophilia*. Cambridge, MA: Harvard Univ. Press, 1984.

Winnicott, D. W. "Communicating and Not Communicating Leading to a Study of Certain Opposites." In *The Maturational Processes and the Facilitating Environment*. New York: International Universities Press, 1965.

Winter, Deborah. *Environmental Psychology*. New York: Harper Collins, 1996.

Wright, Robert. *The Moral Animal: Why We Are the Way We Are: The New Science of Evolutionary Psychology*. New York: Pantheon, 1994.

Young, Iris. *Throwing Like a Girl: And Other Essays in Feminist Philosophy*. Bloomington and London: Indiana Univ. Press, 1990.

Zelinski, Mark. *Outward Bound: The Inward Journey*. Hillsboro, OR: Beyond Words Publishers, 1991.

Zimmerman, Michael, ed. *Environmental Philosophy: From Animal Rights to Radical Ecology*. Englewood Cliffs, NJ: Prentice Hall, 1992. Essays by many leading environmental thinkers.

Zoja, Luigi. *Drugs, Addiction, and Initiation: The Modern Search for Ritual*. M. Romano and R. Mercurio, trans. Boston: Sigo Press, 1989.

Acknowledgments

I am most indebted to the dedicatee, our daughter Rebekah, who gave us thirty-one-and-a-half years of insight, joy, and pride. And to my wife, Donna, a major and ever-growing intellectual stimulus, while continuing her personal support. Then to other adventurers in ideas: Edward Casey held my hand, so to speak, through the roughest of rough drafts of all my work since the 1970s; Glen Mazis gifted me with a brilliant reading of an earlier draft; David Ehrenfeld has helped in more ways than I can say, but at least I can say that; Thomas Berry—a sage presence who guides and emboldens many of us trying to find our way back to the Earth; David Orr, again, a steadying and buoying presence for many; William Barrett, my teacher—dead but still with me somehow. And John Cobb, John Grim, Calvin Martin, Mary-Evelyn Tucker, and John Lachs. I also thank Thelma Tate of the Rutgers University libraries who gave unstintingly of her assistance.

Fred Kayser, a friend since boyhood, has stood by us closely this year. All our friends I thank, as I do my brothers, mother, son, and former sister-in-law, Delores. Also I am indebted to many health practioners. I ask forgiveness of all whom I should have called by name, but have not.

Finally I credit Aina Barten for magnificent editorial guidance. And also personnel at Rowman & Littlefield: Christa Davis Acampora, who acquired this book, and Kermit Hummel, Julie Kirsch, Maureen Mac-Grogan, and Robin Adler who have given it their careful and expert attention all the way. I have needed help and am grateful for what I have received.

Index

❧

About the Author

❧

BRUCE WILSHIRE is professor of philosophy at Rutgers University. He chaired the University College department of philosophy from 1969 to 1979, and the philosophy section of the university from 1977 to 1984. He served as executive secretary of the Society for Phenomenology and Existential Philosophy from 1974 to 1977 and as vice-chair of the Society of Philosophers in America from 1988 to 1993. He has lectured in Europe and Australia. His publications include *William James and Phenomenology: A Study of "The Principles of Psychology"*; *Metaphysics: An Introduction to Philosophy*; *Romanticism and Evolution: The 19th Century*; *Role Playing and Identity: The Limits of Theatre as Metaphor*; and *The Moral Collapse of the University: Professionalism, Purity, Alienation*. A collection of articles has been completed with the working title, *Looking Forward to the First Day: The Story of American Thought—Primal and Pragmatic*. Three of these articles are, "William James, Black Elk, and the Healing Act," "John Dewey: Philosopher and Poet of Nature," "William Ernest Hocking: A Sketch of his Religious-Philosophical Vision." The book will include as an appendix: "The Pluralist Rebellion in the American Philosophical Association, 1978–1985." Wilshire has recently contributed to *The Cambridge Companion to William James* the essay "The Breathtaking Intimacy of the Material World: William James's Last Thoughts."